Fundamentals of Aerospace Engineering

Fundamentals of Aerospace Engineering

Killian Sullivan

CLANRYE INTERNATIONAL
www.clanryeinternational.com

Clanrye International,
750 Third Avenue, 9th Floor,
New York, NY 10017, USA

ISBN: 978-1-64726-124-5

Cataloging-in-Publication Data

Fundamentals of aerospace engineering / Killian Sullivan.
p. cm.
Includes bibliographical references and index.
ISBN 978-1-64726-124-5
1. Aerospace engineering. 2. Aeronautics. 3. Astronautics. I. Sullivan, Killian.
TL545 .F86 2022
629.1--dc23

For information on all Clanrye International publications visit our website at www.clanryeinternational.com

CLANRYE INTERNATIONAL

Contents

Preface

The field of engineering which focuses on the development of spacecraft and aircraft is known as aerospace engineering. This field is closely related to the fields of aeronautical engineering and astronautical engineering. There are a number of areas which are studied under aerospace engineering such as radar cross-section, astrodynamics, avionics and aeroelasticity. Radar-cross-section measures the detectability of objects by radars. The practical problems regarding the motion of spacecraft are handled by astrodynamics through the use of ballistics and celestial mechanics. Avionics deals with the designing and programming of the computer systems in a spacecraft. Aeroelasticity studies the interaction between aerodynamic, elastic and inertial forces. It takes place when an elastic body is exposed to fluid flow. This book is a valuable compilation of topics, ranging from the basic to the most complex advancements in the field of aerospace engineering. It presents the complex subject of aerospace engineering in the most comprehensible and easy to understand language. This book will serve as a valuable source of reference for graduate and postgraduate students.

To facilitate a deeper understanding of the contents of this book a short introduction of every chapter is written below:

Chapter 1- The branch of engineering which is involved in the development of spacecrafts and aircrafts is termed as aerospace engineering. The different aircraft and spacecraft which fall within this discipline are airplanes, helicopters, rockets and satellites. This is an introductory chapter which will introduce briefly all the significant aspects of these aircraft and spacecraft.

Chapter 2- The designing of aircrafts is broadly divided into three states, the conceptual design phase, the preliminary design phase and the detail design phase. The structure of the aircraft consists of different parts such as the cockpit, fuselage, wings, empennage and the flight control surfaces. The topics elaborated in this chapter will help in gaining a better perspective about these aspects of the design and structure of the aircraft.

Chapter 3- The branch of science which focuses on the motion of air and its interaction with solid objects is termed as aerodynamics. Some of the areas of study within aerodynamics in relation to aircraft are effect of wing planform, the angle of attack and the thrust on aircraft. These diverse applications of aerodynamics in aircraft have been thoroughly discussed in this chapter.

Chapter 4- The system which provides the energy for a vehicle to move through outer space or air is termed as a propulsion system. Some of the areas of study related to the propulsion system are the propeller and the aircraft propeller vibration. This chapter has been carefully written to provide an easy understanding of the varied facets of propulsion systems related to aircraft as well as rocket propellants.

Chapter 5- There are numerous instruments and systems which are used in aircraft such as pitot-static systems, compass systems, gyroscopic flight instruments, navigational systems and flight control systems. This chapter discusses in detail these systems and the different instruments which are a part of these systems.

Finally, I would like to thank the entire team involved in the inception of this book for their valuable time and contribution. This book would not have been possible without their efforts. I would also like to thank my friends and family for their constant support.

Killian Sullivan

1

Introduction to Aerospace Engineering

The branch of engineering which is involved in the development of spacecrafts and aircrafts is termed as aerospace engineering. The different aircraft and spacecraft which fall within this discipline are airplanes, helicopters, rockets and satellites. This is an introductory chapter which will introduce briefly all the significant aspects of these aircraft and spacecraft.

Aerospace engineering, also called aeronautical engineering, or astronautical engineering is the field of engineering concerned with the design, development, construction, testing, and operation of vehicles operating in the Earth's atmosphere or in outer space. In 1958 the first definition of aerospace engineering appeared, considering the Earth's atmosphere and the space above it as a single realm for development of flight vehicles. Today the more encompassing aerospace definition has commonly replaced the terms aeronautical engineering and astronautical engineering.

The design of a flight vehicle demands knowledge of many engineering disciplines. It is rare that one person takes on the entire task; instead, most companies have design teams specialized in the sciences of aerodynamics, propulsion systems, structural design, materials, avionics, and stability and control systems. No single design can optimize all of these sciences, but rather there exist compromised designs that incorporate the vehicle specifications, available technology, and economic feasibility.

Aerospace Engineering Functions

The design of a flight vehicle is a complex and time-consuming procedure requiring the integration of many engineering technologies. Supporting teams are formed to provide expertise in these technologies, resulting in a completed design that is the best compromise of all the engineering disciplines. Usually the support teams are supervised by a project engineer or chief designer for technical guidance and by a program manager responsible for program budgets and schedules. Because of the ever-increasing requirement for advanced technology and the high cost and high risk associated with complex flight vehicles, many research and development programs are canceled before completion.

The design process can be dissected into five phases and is the same for most aerospace products. Phase one is a marketing analysis to determine customer specifications or requirements. Aerospace engineers are employed to examine technical, operational, or financial problems. The customer's requirements are established and then passed on to the conceptual design team for the second phase.

The conceptual design team generally consists of aerospace engineers, who make the first sketch attempt to determine the vehicle's size and configuration. Preliminary estimates of the vehicle's performance, weight, and propulsion systems are made. Performance parameters include range, speed, drag, power required, payload, and takeoff and landing distances. This phase may take from a few months to years for major projects.

Phase three is the preliminary design phase. The optimized vehicle design from phase two is used as the starting point. Aerospace engineers perform computer analyses on the configuration; then wind-tunnel models are built and tested. Flight control engineers study dynamic stability and control problems. Propulsion groups supply data necessary for engine selection. Interactions between the engine inlet and vehicle frame are studied. Civil, mechanical, and aerospace engineers analyze the bending loads, stresses, and deflections on the wing, airframe, and other components. Material science engineers aid in selecting low-weight, high-strength materials and may conduct aeroelastic and fatigue tests. Weight engineers make detailed estimates of individual component weights. As certain parameters drive the vehicle design, the preliminary designers are often in close contact with both the conceptual designers and the marketing analysts. The time involved in the preliminary design phase depends on the complexity of the problem but usually takes from six to 24 months.

Phase four, the detailed design phase, involves construction of a prototype. Mechanical engineers, technicians, and draftsmen help lay out the drawings necessary to construct each component. Full-scale mock-ups are built of cardboard, wood, or other inexpensive materials to aid in the subsystem layout. Subsystem components are built and bench-tested and additional wind-tunnel testing is performed. This phase takes from one to three years.

The final phase concerns flight-testing the prototype. Engineers and test pilots work together to assure that the vehicle is safe and performs as expected. If the prototype is a commercial transport aircraft, the vehicle must meet the requirements specified by government organizations such as the Federal Aviation Administration in the United States and the Civil Aviation Authority in the United Kingdom. Prototype testing is usually completed in one year but can take much longer because of unforeseen contingencies. The time required from the perception of a customer's needs to delivery of the product can be as long as 10 to 15 years depending on the complexity of the design, the political climate, and the availability of funding.

High-speed computers have now enabled complex aerospace engineering problems to be analyzed rapidly. More extensive computer programs, many written by aerospace engineers, are being formulated to aid the engineer in designing new configurations.

Branches of Aerospace Engineering

The aerospace engineer is armed with an extensive background suitable for employment in most positions traditionally occupied by mechanical engineers as well as limited positions in the other various engineering disciplines. The transportation, construction, communication, and energy industries provide the most opportunities for non-aerospace applications.

Because land and sea vehicles are designed for optimum speed and efficiency, the aerospace engineer has become a prominent member of the design teams. Because up to half of the power required to propel a vehicle is due to the resistance of the air, the configuration design of low-drag automobiles, trains, and boats offers better speed and fuel economy. The presence of the aerospace engineer in the automobile industry is evident from the streamlined shapes of cars and trucks that evolved during the late 20th century, at a time when gasoline prices were escalating and the aerospace industry was in a lull. Airline companies employ engineers as performance analysts, crash investigators, and consultants. The Federal Aviation Administration makes use of the technical expertise of the aerospace engineer in various capacities.

The construction of large towers, buildings, and bridges requires predictions of aerodynamic forces and the creation of an optimum design to minimize these forces. The consideration of aerodynamic forces of flat surfaces such as the side of a building or superstructure is not new. In 1910 Alexandre-Gustave Eiffel achieved remarkable experimental results measuring the wind resistance of a flat plate, using the Eiffel Tower as a test platform.

Aircraft

An aircraft is a vehicle which can fly. Airplanes, helicopters, and derigeables are all types of aircrafts. They are used to fly through the air, connecting people and places. They are quite important in shortening travel time from place to place, which makes communication and globalization, possible.

Airplane

Airplane, also called aeroplane or plane is any of a class of fixed-wing aircraft that is heavier than air, propelled by a screw propeller or a high-velocity jet, and supported by the dynamic reaction of the air against its wings.

The essential components of an airplane are a wing system to sustain it in flight, tail surfaces to stabilize the wings, movable surfaces to control the attitude of the plane in flight, and a power plant to provide the thrust necessary to push the vehicle through the air. Provision must be made to support the plane when it is at rest on the ground and during takeoff and landing. Most planes feature an enclosed body (fuselage) to house the crew, passengers, and cargo; the cockpit is the area from which the pilot operates the controls and instruments to fly the plane.

Types of Aircraft

There are a number of ways to identify aircraft by type. The primary distinction is between those that are lighter than air and those that are heavier than air.

Lighter-than-air

Aircraft such as balloons, non-rigid airships (blimps), and dirigibles are designed to contain within their structure a sufficient volume that, when filled with a gas lighter than air (heated air, hydrogen, or helium), displaces the surrounding ambient air and floats, just as a cork does on the water. Balloons are not steerable and drift with the wind. Non-rigid airships, which have enjoyed a rebirth of use and interest, do not have a rigid structure but have a defined aerodynamic shape, which contains cells filled with the lifting agent. They have a source of propulsion and can be controlled in all three axes of flight. Dirigibles are no longer in use, but they were lighter-than-air craft with a rigid internal structure, which was usually very large, and they were capable of relatively high speeds.

A colorful airship flies over Dodger Stadium.

Heavier-than-air

This type of aircraft must have a power source to provide the thrust necessary to obtain lift. Simple heavier-than-air craft include kites. These are usually a flat-surfaced structure, often with a stabilizing "tail," attached by a bridle to a string that is held in place on the ground. Lift is provided by the reaction of the string-restrained surface to the wind.

Another type of unmanned aircraft is the unmanned aerial vehicle (UAV). Sometimes called drones or remotely piloted vehicles (RPVs), these aircraft are radio-controlled from the air or the ground and are used for scientific and military purposes.

Israeli Aircraft Industries Searcher, a reconnaissance unmanned aerial vehicle.

Unpowered manned heavier-than-air vehicles must be launched to obtain lift. These include hang gliders, gliders, and sailplanes.

General Atomics MQ-1 Predator, a reconnaissance unmanned aerial vehicle.

Hang gliders are aircraft of various configurations in which the pilot is suspended beneath the (usually fabric) wing to provide stability and control. They are normally launched from a high point. In the hands of an experienced pilot, hang gliders are capable of soaring (using rising air columns to obtain upward gliding movement).

Glider.

Gliders are usually used for flight training and have the capability to fly reasonable distances when they are catapulted or towed into the air, but they lack the dynamic sophistication of sailplanes. These sophisticated unpowered craft have wings of unusually high aspect ratio (that is, a long wing span in proportion to wing width). Most sailplanes are towed to launch altitude, although some employ small, retractable auxiliary engines. They are able to use thermals (currents more buoyant than the surrounding air, usually caused by higher temperature) and orographic lift to climb to higher altitude and to glide for great distances. Orographic lift results from the mechanical effect of wind blowing against a terrain feature such as a cliff. The force of the wind is deflected upward by the face of the terrain, resulting in a rising current of air.

Ultralights, which were originally merely hang gliders adapted for power by the installation of small engines similar to those used in chain saws, have matured into specially designed aircraft of very low weight and power but with flying qualities similar to conventional light aircraft. They are intended primarily for pleasure flying, although advanced models are now used for training, police patrol, and other work, including a proposed use in combat.

Whooping cranes following an ultralight aircraft.

Experimental craft have been designed to make use of human and solar power. These are very lightweight, sophisticated aircraft, designed with heavy reliance on computers and using the most modern materials. Paul MacCready of Pasadena, California, U.S., was the leading exponent of the discipline; he first achieved fame with the human-powered Gossamer Condor, which navigated a short course in 1977. Two of his later designs, the human-powered Gossamer Albatross and the solar-powered Solar Challenger, successfully crossed the English Channel. Others in the field have carried on MacCready's work, and a human-powered helicopter has been flown. Solar-powered aircraft are similar to human-powered types, except that they use solar panels to convert the Sun's energy directly to power an electric motor.

Civil Aircraft

All nonmilitary planes are civil aircraft. These include private and business planes and commercial airliners.

Private aircraft are personal planes used for pleasure flying, often single-engine monoplanes with non-retractable landing gear. They can be very sophisticated, however, and may include such variants as: "warbirds," ex-military planes flown for reasons of nostalgia, ranging from primary trainers to large bombers; "homebuilts," aircraft built from scratch or from kits by the owner and ranging from simple adaptations of Piper Cubs to high-speed, streamlined four-passenger transports; antiques and classics, restored older aircraft flown, like the warbirds, for reasons of affection and nostalgia; and aerobatic planes, designed to be highly maneuverable and to perform in air shows.

A Beechcraft Baron airplane coming in for a landing:
The aircraft is used mainly for private and corporate travel.

Business aircraft are used to generate revenues for their owners and include everything from small single-engine aircraft used for pilot training or to transport small packages over short distances to four-engine executive jets that can span continents and oceans. Business planes are used by salespeople, prospectors, farmers, doctors, missionaries, and many others. Their primary purpose is to make the best use of top executives' time by freeing them from airline schedules and airport operations. They also serve as an executive perquisite and as a sophisticated inducement for potential customers. Other business aircraft include those used for agricultural operations, traffic reporting, forest-fire fighting, medical evacuation, pipeline surveillance, freight hauling, and many other applications. One unfortunate but rapidly expanding segment of the business aircraft population is that which employs aircraft illegally for transporting narcotics and other illicit drugs. A wide variety of similar aircraft are used for specialized purposes, like the investigation of thunderstorms, hurricane tracking, aerodynamic research and development, engine testing, high-altitude surveillance, advertising, and police work.

Gulfstream G450: The Gulfstream G450 business jet, capable of carrying eight passengers up to 5,000 miles (8,000 km).

Commercial airliners are used to haul passengers and freight on a scheduled basis between selected airports. They range in size from single-engine freight carriers to the Boeing 747 and in speed from below 200 miles per hour to supersonic, in the case of the Anglo-French Concorde, which was in service from 1976 to 2003.

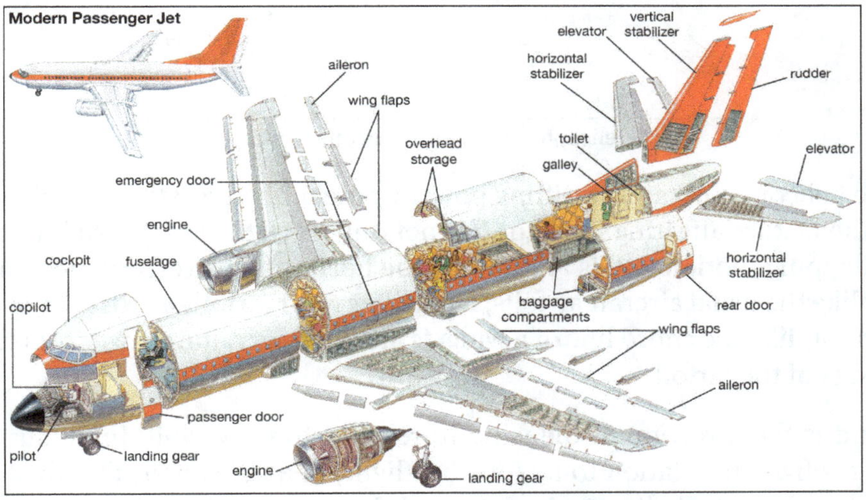

Boeing 737: Cross sections of a Boeing 737 passenger jet.

Helicopter

Helicopter is an aircraft with one or more power-driven horizontal propellers or rotors that enable it to take off and land vertically, to move in any direction, or to remain stationary in the air. Other vertical-flight craft include autogiros, convertiplanes, and V/STOL aircraft of a number of configurations.

The idea of taking off vertically, making the transition to horizontal flight to the destination, and landing vertically has been for centuries the dream of inventors. It is the most logical form of flight, dispensing as it does with large landing fields located far from city centres and the inevitable intervening modes of travel—automobile, subway, bus—that flight in conventional aircraft usually requires. But vertical flight is also the most demanding challenge in flying, requiring more sophistication in structure, power, and control than conventional fixed-wing aircraft.

Principles of Flight and Operation

Unlike fixed-wing aircraft, the helicopter's main airfoil is the rotating blade assembly (rotor) mounted atop its fuselage on a hinged shaft (mast) connected with the vehicle's engine and flight controls. In comparison to airplanes, the tail of a helicopter is somewhat elongated and the rudder smaller; the tail is fitted with a small antitorque rotor (tail rotor). The landing gear sometimes consists of a pair of skids rather than wheel assemblies.

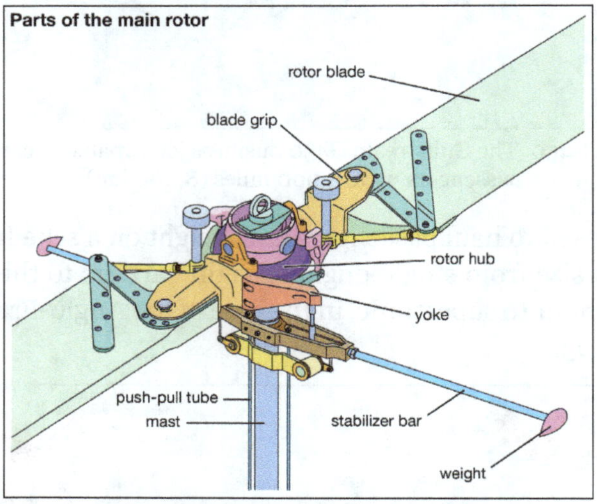

Helicopter: Parts of a helicopter rotor.

The fact that the helicopter obtains its lifting power by means of a rotating airfoil (the rotor) greatly complicates the factors affecting its flight, for not only does the rotor turn but it also moves up and down in a flapping motion and is affected by the horizontal or vertical movement of the helicopter itself. Unlike the usual aircraft airfoils, helicopter rotor airfoils are usually symmetrical. The chord line of a rotor, like the chord line of a wing, is an imaginary line drawn from the leading edge to the trailing edge of the airfoil.

The relative wind is the direction of the wind in relation to the airfoil. In an airplane, the flight path of the wing is fixed in relation to its forward flight; in a helicopter, the flight path of the rotor advances forward (to the helicopter's nose) and then rearward (to the helicopter's tail) in the process of its circular movement. Relative wind is always considered to be in parallel and opposite direction to the flight path. In considering helicopter flight, the relative wind can be affected by the rotation of the blades, the horizontal movement of the helicopter, the flapping of the rotor blades, and wind speed and direction. In flight, the relative wind is a combination of the rotation of the rotor blade and the movement of the helicopter.

Like a propeller, the rotor has a pitch angle, which is the angle between the horizontal plane of rotation of the rotor disc and the chord line of the airfoil. The pilot uses the collective and cyclic pitch control to vary this pitch angle. In a fixed-wing aircraft, the angle of attack (the angle of the wing in relation to the relative wind) is important in determining lift. The same is true in a helicopter, where the angle of attack is the angle at which the relative wind meets the chord line of the rotor blade.

Angle of attack and pitch angle are two distinct conditions. Varying the pitch angle of a rotor blade changes its angle of attack and hence its lift. A higher pitch angle (up to the point of stall) will increase lift; a lower pitch angle will decrease it. Individual blades of a rotor have their pitch angles adjusted individually.

Rotor speed also controls lift—the higher the revolutions per minute (rpm), the higher the lift. However, the pilot will generally attempt to maintain a constant rotor rpm and will change the lift force by varying the angle of attack.

As with fixed-wing aircraft, air density (the result of air temperature, humidity, and pressure) affects helicopter performance. The higher the density, the more lift will be generated; the lower the density, the less lift will be generated. Just as in fixed-wing aircraft, a change in lift also results in a change in drag. When lift is increased by enlarging the angle of pitch and thus the angle of attack, drag will increase and slow down the rotor rpm. Additional power will then be required to sustain a desired rpm. Thus, while a helicopter is affected like a conventional aircraft by the forces of lift, thrust, weight, and drag, its mode of flight induces additional effects.

In a helicopter, the total lift and thrust forces generated by the rotor are exerted perpendicular to its plane of rotation. When a helicopter hovers in a windless condition, the plane of rotation of the rotor (the tip-path plane) is parallel to the ground and the sum of the weight and drag forces are exactly balanced by the sum of the thrust and lift forces. In vertical flight, the components of weight and drag are combined in a single vector that is directed straight down; the components of lift and thrust are combined in a single vector that is directed straight up. To achieve forward flight in a helicopter, the plane of rotation of the rotor is tipped forward. (It should be understood that the helicopter's rotor mast does not tip but rather the individual rotor blades within the plane of rotation have their pitch angle varied.) For sideward flight, the plane of the rotation of the rotor is tilted in the direction desired. For rearward flight, the plane of the rotation of the rotor is tilted rearward.

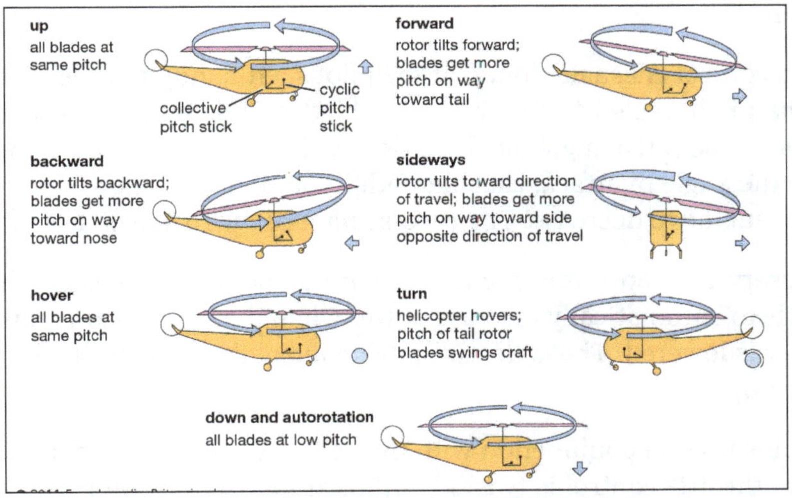

Helicopter: Controlling the direction of flight in a helicopter.

Because the rotor is powered, there is an equal and opposite torque reaction, which tends to rotate the fuselage in a direction opposite to the rotor. This torque is offset by the tail rotor (antitorque rotor) located at the end of the fuselage. The pilot controls the thrust of the tail rotor by means of foot pedals, neutralizing torque as required.

There are other forces acting upon a helicopter not found in a conventional aircraft. These include the gyroscopic precession effect of the rotor—that is, the dissymmetry of lift created by the forward movement of the helicopter, resulting in the advancing blade having more lift and the retreating blade less. This occurs because the advancing blade has a combined speed of the blade velocity and the speed of the helicopter in forward flight, while the retreating blade has the difference between the blade velocity and the speed of the helicopter. This difference in speed causes a difference in lift—the advancing blade is moving faster and hence is generating more lift. If uncontrolled, this would result in the helicopter rolling. However, the difference in lift is compensated for by the blade flapping and by cyclic feathering (changing the angle of pitch). Because the blades are attached to a rotor hub by horizontal flapping hinges, which permit their movement in a vertical plane, the advancing blade flaps up, decreasing its angle of attack, while the retreating blade flaps down, increasing its angle of attack. This combination of effects equalizes the lift. (Blades also are attached to the hub by a vertical hinge, which permits each blade to move back and forth in the plane of rotation. The vertical hinge dampens out vibration and absorbs the effect of acceleration or deceleration.) In addition, in forward flight, the position of the cyclic pitch control causes a similar effect, contributing to the equalization of lift.

Other forces acting upon helicopters include coning, the upward bending effect on blades caused by centrifugal force; Coriolis effect, the acceleration or deceleration of the blades caused by the flapping movement bringing them closer to (acceleration) or farther away from (deceleration) the axis of rotation; and drift, the tendency of the tail rotor thrust to move the helicopter in hover.

Control Functions

A helicopter has four controls: collective pitch control, throttle control, anti-torque control, and cyclic pitch control.

The collective pitch control is usually found at the pilot's left hand; it is a lever that moves up and down to change the pitch angle of the main rotor blades. Raising or lowering the pitch control increases or decreases the pitch angle on all blades by the same amount. An increase in the pitch angle will increase the angle of attack, causing both lift and drag to increase and causing the rpm of the rotor and the engine to decrease. The reverse happens with a decrease in pitch angle.

Because it is necessary to keep rotor rpm as constant as possible, the collective pitch control is linked to the throttle to automatically increase power when the pitch lever is raised and decrease it when the pitch lever is lowered. The collective pitch control thus acts as the primary control both for altitude and for power.

The throttle control is used in conjunction with the collective pitch control and is an integral part of its assembly. The throttle control is twisted outboard to increase rotor rpm and inboard to decrease rpm.

The antitorque controls are pedals linked to operate a pitch change mechanism in the tail rotor gearbox. A change in pedal position changes the pitch angle of the tail rotor to offset torque. As torque varies with every change of flight condition, the pilot is required to change pedal position accordingly. The antitorque control does not control the direction of flight.

It was stated that the lift/thrust force is always perpendicular to the plane of rotation of the rotor.

The cyclic pitch control, a stick-type control found to the pilot's right, controls the direction of flight by tipping the plane of rotation in the desired direction. The term cyclic derives from the sequential way each blade's pitch is changed so that it takes the flight path necessary to effect the change in direction.

Differences in Helicopter and Airplane Design and Construction

The most immediate and obvious difference in the construction of a fixed-wing aircraft and a helicopter is of course the latter's use of a rotor instead of a wing. There are many other critical additions, however, including the use of a tail rotor to offset torque. (Some helicopters use a "no tail rotor" system, in which low-pressure air is circulated through a tail boom to control the torque of the spinning main rotor.) Less obvious are such additions as the transmission system, which is used to transfer power from the engine to the rotor, tail rotor, and other accessories; the clutch, used to engage the engine and transmission with the rotor; and the mechanics of the rotor system itself.

Helicopter: A helicopter in flight.

The first helicopters were quite primitive, with skids instead of wheeled landing gear, open cockpits, and unaired fuselage sections. Helicopters are now as fully equipped as airplanes, with retractable landing gear and full instrumentation and navigation equipment, and are provided with whatever accoutrements may be necessary to accomplish the specific task at hand. For example, some helicopters are flying ambulances, especially equipped with a complete set of intensive-care accessories. Others function as electronic news gatherers, with appropriate sensors and telecommunications equipment.

The design and operation of helicopters have derived the same advances from computers and composites as have other aircraft, especially in the design and construction of the rotor blades. One of the more important improvements is in the simplification of flight-control systems, where a simple side stick controller, with the assistance of computers, performs the functions of the collective, cyclic, and throttles controls.

Helicopter designs have included a number of optional rotor configurations, such as rotors that stop to serve as a fixed wing for forward flight; rotors that fold in a stream-wise direction to blend in with, or be stowed within, the fuselage contours, lift being provided by a stub wing; and X-shaped rotors that rotate for takeoff and landing but are fixed for lift in flight.

In sum, the additional forces imposed upon a helicopter by its very concept delayed its development, made it relatively more difficult to control than fixed-wing aircraft, and, in general, impeded its use. While generally considered more expensive to operate than conventional fixed-wing aircraft, a true comparison of costs cannot be made without assessing the additional advantages conferred by the vertical flight capability. The popularity of the helicopter indicates that users willingly pay any additional costs involved to obtain that capability. In some applications—medical evacuation, supplying of oil drilling rigs, spreading of certain agricultural agents, to name but a few—it is irreplaceable.

Spacecraft

A spacecraft is a vehicle designed for travel or operation in space beyond the earth's atmosphere or in orbit around the earth.

BepiColombo's cruise configuration.

A spacecraft has a number of essential components, such as an engine, power subsystem, steering system and communications system, in addition to science instruments. Most of these systems are housed in a section called the service module, while the science instruments make up the payload module. These are encased inside the spacecraft's main structural unit and connected by a 'harness', the electrical framework on which the spacecraft electronics systems communicate.

Propulsion and Thrust

The propulsion system that a satellite carries is dictated by the orbital position that it needs to reach and by the type of mission. Additionally to the launch vehicle-provided propulsion, there may be an upper stage (e.g. Fregat) that delivers the spacecraft to a higher or to an escape orbit. The spacecraft itself may carry its own (attached or jettisonable) propulsion module for orbital manoeuvres or orbit adjustment. Finally, smaller thrusters may be included for fine orbit adjustments and orbit maintenance, the so-called station keeping.

'Cold gas' propulsion will be used by Gaia and Lisa Pathfinder and is the simplest type of chemical propulsion. The propulsion system consists of pressurized gas and a nozzle, and whenever thrust is needed the nozzle is opened and some of the gas is expelled.

Most satellites also have simple hydrazine thrusters and reaction or momentum wheels to control attitude and orbit.

Solar-electric propulsion using ion engines is an alternative way to travel in space. Ion engines shoot out a spray of ionized particles to deliver their thrust to push the spacecraft on its way. Ion engines typically get the high power required to ionize the propellant particles from the spacecraft's solar panels. They work their magic in a leisurely way and can go on pushing gently for months or even years, for as long as the Sun shines and the small supply of propellant lasts.

Power

A spacecraft must generate and distribute adequate power to operate its subsystems and instruments. The power generation is usually catered for by solar panels, which convert sunlight into electricity. When the spacecraft is in eclipse, that is in the shadow of a celestial object or turned away from the Sun and the panels thus do not generate power, batteries kick-in. Depending on the type of spacecraft, solar panels may take on a traditional winged shape (for three-axis stabilized satellites) or be wrapped around the cylindrical body of a spinning spacecraft.

Different spacecraft units may require different voltages. Some spacecraft have central units to convert and stabilise the 'raw' voltage produced by the solar array before it is distributed, while for others the conversion happens within the units.

Navigation

A navigation system, usually referred to as 'Attitude and Orbit Control' is essential for the spacecraft to maintain its attitude, so that it can be directed towards its intended destination. This capability is also required to keep its solar panels pointed towards the Sun and its instruments pointing at specific targets. Attitude control is achieved with electronic 'eyes' (sun and star sensors). These can sense the location of the Sun and track the stars allowing the spacecraft's position and orientation to be calculated. A propulsion system or reaction wheels controlled by appropriate software ensure then that the spacecraft is moved or oriented appropriately. Most modern spacecraft are 'three-axis stabilised' so that they sit 'upright' in space, while other satellites may spin to achieve stability. For example, Planck revolves about a Sun-pointing axis once per minute to stabilise its attitude.

Planck spinning in space.

Communications and Data Handling

The communications system returns science data, that is passed on to the scientists, and 'housekeeping' data, called telemetry, which allows engineers back on Earth to monitor the condition of the spacecraft. The on-board communications system also receives all incoming commands. Communications may be carried out at different frequencies using a combination of dishes and high- or low-gain antennas.

Instructions are received from the ground in the form of individual commands or sequences to be executed at pre-defined times. The commands are processed by the data handling system that either executes them immediately or stores them for later execution.

The data handling system comprises processors, on-board memory, and analog to digital and digital to analog conversion systems.

Rocket

A rocket is a vehicle, missile, or aircraft that obtains thrust by reaction to the ejection of fast-moving fluid from within a rocket engine. The study and development of rockets is called rocketry.

A Soyuz rocket, at Baikonur launch pad.

In chemically powered rockets, the combustion of propellant produces hot gases that act against the inside of combustion chambers and expansion nozzles. In this process, the gases are accelerated to extremely high speeds, and, because every action has an equal and opposite reaction, generate a large thrust on the rocket. Thus, rockets contain a large amount of energy in an easily liberated form, and they can be very dangerous. However, the risks can be minimized through careful design, testing, and construction.

Rockets are used for fireworks and weaponry, as launch vehicles for artificial satellites, and for human spaceflight and exploration of other planets. Compared to other propulsion systems, they are very lightweight, enormously powerful, and can achieve extremely high speeds.

Types

There are many different types of rockets, and a comprehensive list can be found in rocket engine—they range in size from tiny models such as water rockets or small solid rockets that can be purchased at a hobby store, to the enormous Saturn V used for the Apollo program.

Rockets at minimum consist of propellant, one or more rocket engines, stabilization device(s) and a structure (typically monocoque) to hold these components together. Many rockets also have an aerodynamic fairing such as a nose cone.

Saturn V is the biggest rocket to have successfully flown.

Most current rockets are chemically powered rockets (internal combustion engines) that emit a hot exhaust gas. A chemical rocket engine can use gas propellant, solid propellant, liquid propellant, or a hybrid mixture of both solid and liquid. A chemical reaction is initiated between the fuel and the oxidizer in the combustion chamber, and the resultant hot gases accelerate out of a nozzle (or nozzles) at the rearward-facing end of the rocket. The acceleration of these gases through the engine exerts force ("thrust") on the combustion chamber and nozzle, propelling the vehicle (in accordance with Newton's Third Law).

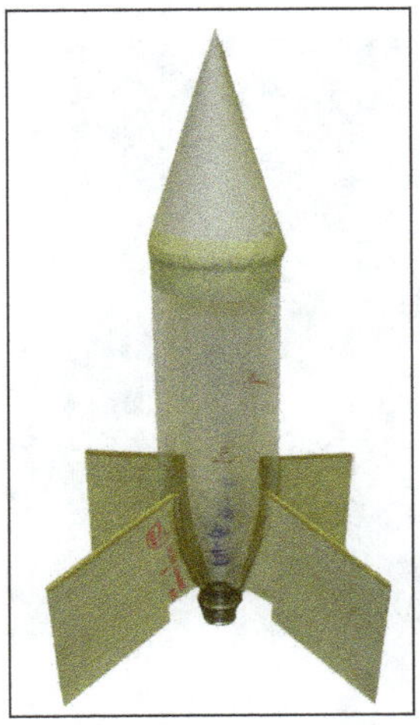

Water rockets are flown for recreational purposes.

Not all rockets use chemical reactions. Steam rockets, for example, release superheated water through a nozzle where it instantly flashes to high velocity steam, propelling the rocket. The efficiency of steam as a rocket propellant is relatively low, but it is simple and reasonably safe, and the propellant is cheap and widely available. There are even proposals to use steam rockets for interplanetary transport using either nuclear or solar heating as the power source to vaporize water collected from around the solar system.

Rockets where the heat is supplied from other than the propellant, such as steam rockets, are classed as external combustion engines. Other examples of external combustion rocket engines include most designs for nuclear powered rocket engines. Use of hydrogen as the propellant for external combustion engines gives very high velocities.

Uses

Weaponry

A Boeing MIM-115 surface-to-air missile.

In many military weapons, rockets are used to propel payloads to their targets. A rocket and its payload together are generally referred to as a missile, especially when the weapon has a guidance system.

Science

A Bumper sounding rocket.

Sounding rockets are commonly used to carry instruments that take readings from 50 kilometers (30 mi) to 1,500 kilometers (930 mi) above the surface of the Earth, the altitudes between those reachable by weather balloons and satellites.

Launch

Due to their high exhaust velocity (Mach ~10+), rockets are particularly useful when very high speeds are required, such as orbital speed (Mach 25+). Indeed, rockets remain the only way to launch spacecraft into orbit. They are also used to rapidly accelerate spacecraft when they change orbits or de-orbit for landing. Also, a rocket may be used to soften a hard parachute landing immediately before touchdown. Spacecraft delivered into orbital trajectories become artificial satellites.

Space Shuttle Atlantis burning for orbit.

Hobby and Entertainment

Hobbyists build and fly Model rockets of various types and rockets are used to launch both commercially available fireworks and professional fireworks displays.

Physics

Operation

In all rockets, the exhaust is formed from propellants carried within the rocket prior to use. Rocket thrust is due to the rocket engine, which propels the rocket forwards by expelling the exhaust rearwards at extreme high speed.

In a closed chamber, the pressures are equal in each direction and no acceleration occurs. If an opening is provided at the bottom of the chamber then the pressure is no longer acting on that side. The remaining pressures gives a resultant thrust in the side opposite the opening which provides thrust. Using a nozzle increases the forces further, in fact multiplies the thrust depending on the area ratio of the nozzle.

If propellant gas is continuously added to the chamber then this disequilibrium of pressures can be maintained for as long as propellant remains.

As the remaining propellant decreases, the vehicle's acceleration tends to increase until it runs out of propellant, and this means that much of the speed change occurs towards the end of the burn when the vehicle is much lighter.

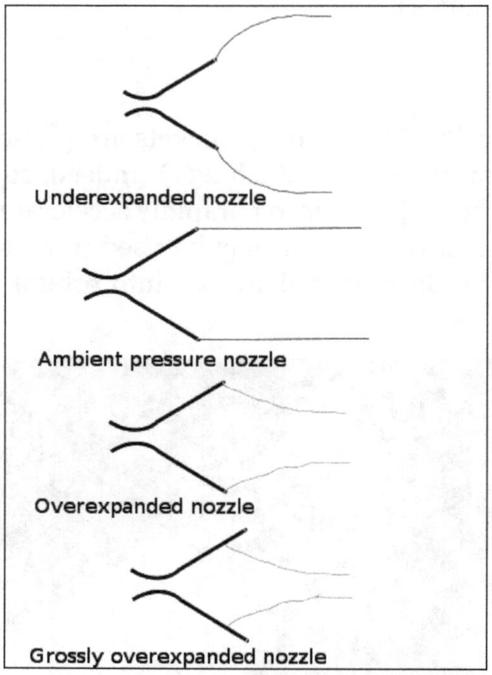

Figure indicates, if the pressure of the exhaust jet varies from atmospheric pressure, nozzles can be said to be underexpanded, ambient or overexpanded. If under or overexpanded then loss of efficiency occurs, grossly overexpanded nozzles lose less efficiency, but the exhaust jet is usually unstable. Rockets become progressively more underexpanded as they gain altitude. Note that almost all rocket engines will be momentarily grossly underexpanded during startup in an atmosphere.

Net Thrust

Below is an approximate equation for calculating the gross thrust of a rocket:

$$F_g = \dot{m}\, V_e + A_e(P_e - P_{amb})$$

where,

$\dot{m} =$ exhaust gas mass flow,

$V_e =$ jet velocity at nozzle exit plane,

$A_e =$ flow area at nozzle exit plane,

$P_e =$ static pressure at nozzle exit plane,

$P_{amb} =$ ambient (or atmospheric) pressure.

Since, unlike a jet engine, a conventional rocket motor lacks an air intake, there is no 'ram drag' to deduct from the gross thrust. Consequently the net thrust of a rocket motor is equal the gross thrust.

The $\dot{m} V_e$ term represents the momentum thrust, which remains constant at a given throttle setting, whereas the $A_e(P_e - P_{amb})$ term represents the pressure thrust term. At full throttle, the net thrust of a rocket motor improves slightly with increasing altitude, because the reducing atmospheric pressure increases the pressure thrust term.

Note that because rockets choke at the throat, the pressure at the exit is ideally exactly proportional to the propellant flow $\dot{m}$, provided the mixture ratios and combustion efficiencies are maintained. It is thus quite usual to rearrange the above equation slightly:

$$F_g = \dot{m}.V_{e(vac)} - A_e P_{amb}$$

Where,

$V_{e(vac)}$ = the effective exhaust velocity in a vacuum of that particular engine.

Mass Ratios

Mass ratio is the ratio between fully fuelled mass and the mass when the usable fuel has all been exhausted. A high mass ratio is desirable for good performance, since it indicates that the rocket is lightweight and hence performs better, for essentially the same reasons that low weight is desirable in sports cars.

Rockets as a group have the highest thrust-to-weight ratio of any type of engine; and this helps vehicles achieve high mass ratios, which improves the performance of flights. The higher this ratio, the fewer engines mass is needed to be carried and permits the carrying of even more propellant, this enormously improves performance.

Achievable mass ratios are highly dependent on many factors such as the type of engine the vehicle uses and structural safety margins. Common mass ratios for launch vehicles are 20:1 for dense propellants such as liquid oxygen and kerosene, 25:1 for dense monopropellants such as hydrogen peroxide, and 10:1 or worse for liquid oxygen and liquid hydrogen.

Delta-v (Rocket Equation)

The delta-v capacity of a rocket is the theoretical total change in velocity that a rocket can achieve without any external interference (without air drag or gravity or other forces).

The speeds that a rocket vehicle can reach can be calculated by the Tsiolkovsky rocket equation, which gives the speed difference ("delta-v") in terms of the exhaust speed and ratio of initial mass to final mass ("mass ratio").

Energy Efficiency

At take-off the rocket has a great deal of energy in the form of fuel and oxidiser stored within the vehicle, and it is of course desirable that as much of the energy stored in the propellant ends up as kinetic or potential energy of the body of the rocket as possible.

Energy from the fuel is lost in air drag and is used to gain altitude. However, much of the lost energy ends up in the exhaust.

One hundred percent efficiency within the engine (η_c) would mean that all of the heat energy of the combustion products is converted into kinetic energy of the jet. This is not possible, but nozzles come surprisingly close: When the nozzle expands the gas, the gas is cooled and accelerated, and an energy efficiency of up to 70 percent can be achieved. Most of the rest is heat energy in the exhaust that is not recovered. This compares very well with other engine designs. The high efficiency is a consequence of the fact that rocket combustion can be performed at very high temperatures and the gas is finally released at much lower temperatures, and so giving good Carnot efficiency.

However, engine efficiency is not the whole story. In common with many jet-based engines, but particularly in rockets due to their high and typically fixed exhaust speeds, rocket vehicles are extremely inefficient at low speeds irrespective of the engine efficiency. The problem is that at low speeds, the exhaust carries away a huge amount of kinetic energy rearward.

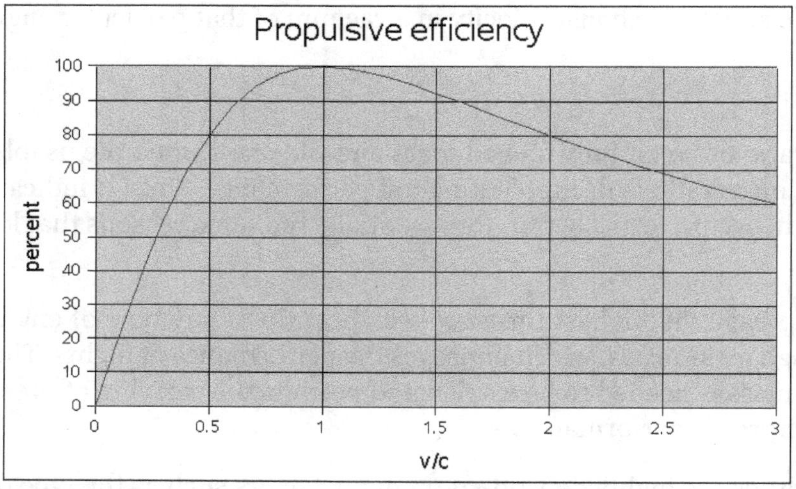

Rocket propulsive efficiency as a function of vehicle speed divided by effective exhaust speed.

However as speeds rise, the resultant exhaust speed goes down, and the overall vehicle energetic efficiency rises, reaching a peak of (theoretically) 100 percent of the engine efficiency when the vehicle is traveling exactly at the same speed that the exhaust is emitted; and then the exhaust in principle stops dead in space behind the moving vehicle. The efficiency then drops off again at even higher speeds as the exhaust ends up traveling forwards behind the vehicle.

$$\eta_p = \frac{2\frac{u}{c}}{1+\left(\frac{u}{c}\right)^2}$$

Since the energy ultimately comes from fuel, these joint considerations mean that rockets are mainly useful when a very high speed is required, and thus they are rarely if ever used for general aviation. Jet engines which have a better match between speed and jet exhaust speed such as turbofans dominate for subsonic and supersonic atmospheric use while rockets work best at hypersonic speeds. On the other hand rockets do also see many short-range *relatively* low speed military applications where their low-speed inefficiency is outweighed by their extremely high thrust and hence high accelerations.

Staging

Often, the required velocity (delta-v) for a mission is unattainable by any single rocket because the propellant, structure, guidance, and engines take a particular minimum percentage of take-off mass.

The mass ratios that can be achieved with a single set of fixed rocket engines and tankage varies depends on acceleration required, construction materials, tank layout, engine type and propellants used, but for example the first stage of the Saturn V, carrying the weight of the upper stages, was able to achieve a mass ratio of about 10.

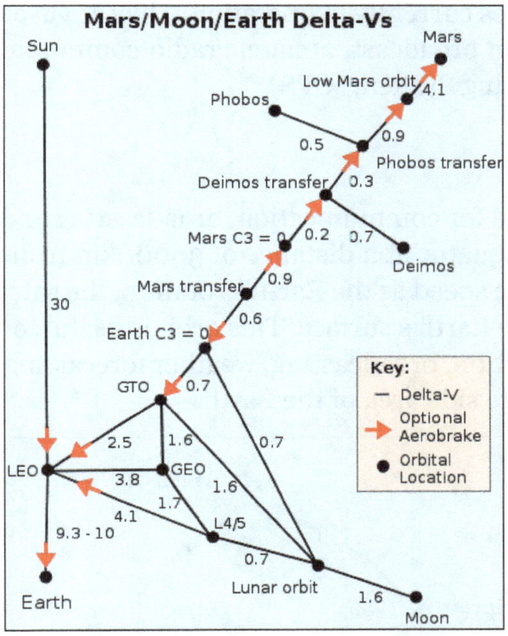

A map of approximate Delta-v's around the solar system between Earth and mas.

This problem is frequently solved by staging—the rocket sheds excess weight (usually tankage and engines) during launch to reduce its weight and effectively increase its mass ratio. Staging is either *serial* where the rockets light one after the previous stage has fallen away, or *parallel*, where rockets are burning together and then detach when they burn out.

Apollo 6 while dropping the interstage ring.

Typically, the acceleration of a rocket increases with time (if the thrust stays the same) as the weight of the rocket decreases as propellant is burned. Discontinuities in acceleration will occur when stages burn out, often starting at a lower acceleration with each new stage firing.

Satellite

A satellite is an artificial body placed in orbit round the earth or moon or another planet in order to collect information or for communication. In a communications context, a satellite is a specialized wireless receiver/transmitter that is launched by a rocket and placed in orbit around the earth. There are hundreds of satellites currently in operation. They are used for such diverse purposes as weather forecasting, television broadcast, amateur radio communications, Internet communications, and the Global Positioning System, (GPS).

Satellite Orbits

The geostationary orbit is used for communication, broadcasting and weather satellites. This orbit takes a position around the equator at a distance of 36000km from the Earth's surface. Because the satellite moves at the same speed as the Earth's rotation, the satellite maintains the same position over a certain point of the Earth's surface. This orbit is useful for "serving at all times in a certain area" such as communication, broadcasting, weather forecasting missions. But these satellites cannot be used on the opposite side area of the Earth.

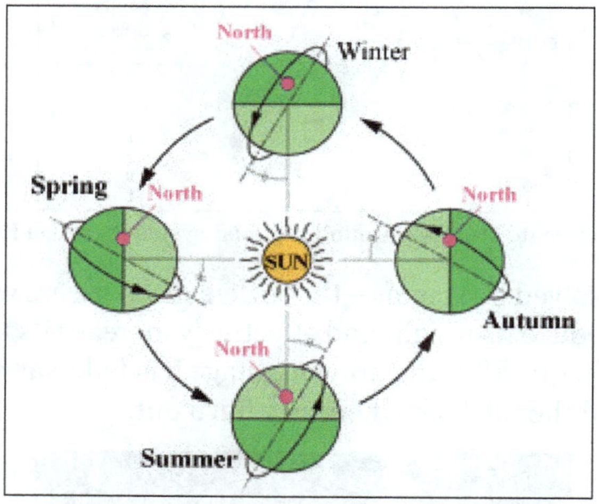

Sun-synchronous Quasi-periodical Orbit

This orbit is made by combining the synchronous orbit and the quasi periodical orbit, and many Earth observation satellites employ this orbit. The satellite on this orbit passes by a certain place of the Earth's surface at the same local solar time (the orbit plane keeps a constant solar angle), so the satellite observes the same sun light reflection from the Earth's surface. This enables the observed images of each place to be easily compared and analyzed with the same sun light reflection phenomena. The satellite on this orbit observes all Earth surfaces, so they are utilized in all areas of the world. Also, the satellite on this orbit has a regression cycle in which the satellite passes the same place after respective pass cycles. Then this satellite observes the certain place with the same local solar time and same orbit position for each respective pass cycles.

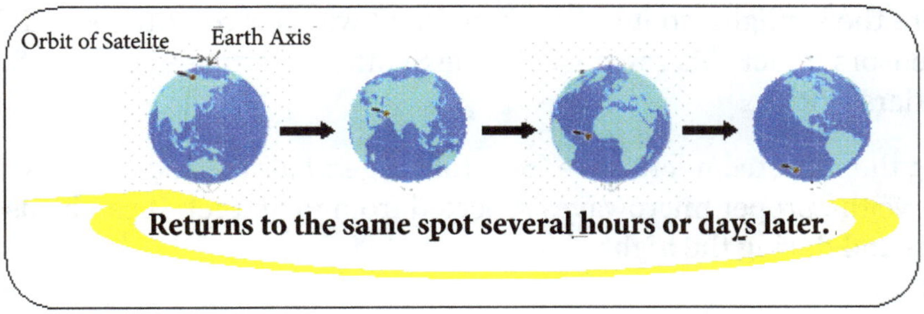

Other Orbits

The Tropical Rainfall Measuring Mission (TRMM) satellite is assigned to an orbit that observes all tropical regions along the equator at various local solar times. The orbit is selected for the purpose of its mission.

Satellite Remote Sensing

Earth observation satellites are the satellites equipped with remote sensing technology. Various sensors are equipped on these satellites in accordance with their respective missions.

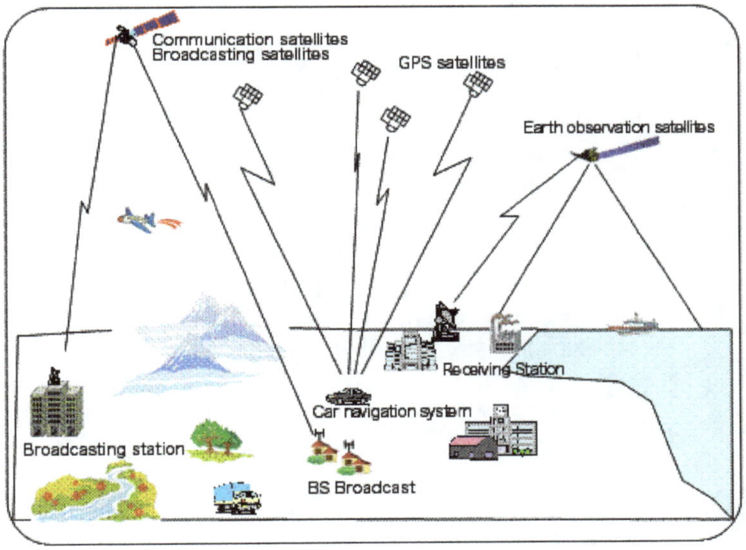

Typical Sensors

The sensors are categorized into three kinds:

- Observing the reflection of sunlight or the Earth's radiation: Optical sensors these sensors observe the reflection of sunlight including invisible ultraviolet and near, medium, thermal infrared rays, and the Earth's radiation.
- Transmitting microwaves to the target and receiving the reflected microwave from the target: Active microwave sensors, these sensors transmit microwaves to the target and receive the reflected microwaves from the target. This method is not affected by clouds and does

not require the sunlight, so it can be used in all weather conditions and during the night. Typical sensors under this category are the Synthetic Aperture Raders (SAR) and Precipitation Radars (PR).

- Observing the reflected microwave from the target: Passive microwave sensors, these sensors observe the proper microwaves reflected from the target. It can be used in all weather conditions and during the night.

References

- Aerospace-engineering, technology: britannica.com, Retrieved 31 March, 2019
- Airplane, technology: britannica.com, Retrieved 14 July, 2019
- Helicopter, technology: britannica.com, Retrieved 17 May, 2019
- Anatomy-of-a-spacecraft, space-science, our-activities: esa.int, Retrieved 19 April, 2019
- Rocket, entry: newworldencyclopedia.org, Retrieved 5 February, 2019
- Satellite, definition: searchmobilecomputing.techtarget.com, Retrieved 26 July, 2019
- Sensing-2, sensing, knowledge: restec.or.jp, Retrieved 21 May, 2019

2

Aircraft: Design and Structure

The designing of aircrafts is broadly divided into three states, the conceptual design phase, the preliminary design phase and the detail design phase. The structure of the aircraft consists of different parts such as the cockpit, fuselage, wings, empennage and the flight control surfaces. The topics elaborated in this chapter will help in gaining a better perspective about these aspects of the design and structure of the aircraft.

Aircraft Design

Aircraft design focuses on the integration of new technologies and systems with current and advanced configurations. This includes new structures, materials and manufacturing processes.

Three Stages of Aircraft Design

Conceptual Design

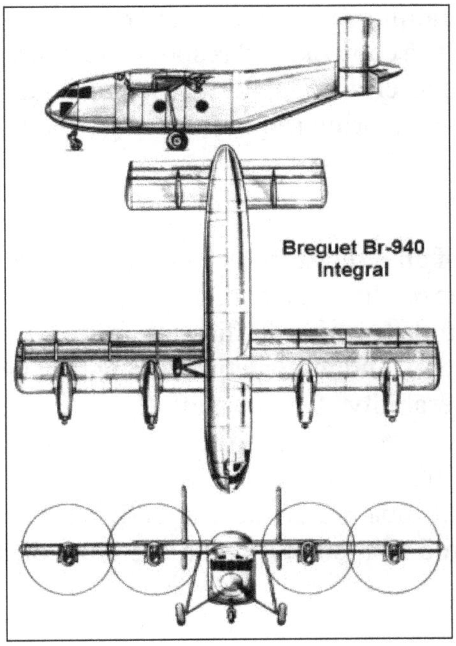

The first stage of aircraft design is conceptual design. This is the very beginning of the design process in which rough sketches are created with respect to the aircraft's configurations. Designers seek to achieve a design that meets all of the aircraft's requirements, including aerodynamics, propulsion, performance, structural systems, control systems and more. Designers must also consider elements such as the shape of the aircraft's fuselage, wing location, engine size and more.

Preliminary Design

After completing the conceptual design, the next phase is preliminary design. During this stage, the conceptual design is optimized to fit into the necessary parameters. Engineers may use the existing designs to conduct wind tunnel testing and fluid dynamic calculations. Furthermore, structural and control analyses are performed during this stage. Engineers will also look for structural defects and flaws, correcting them before proceeding to the third and final stage of the design process.

Detail Design

Last but not least is the detail design phase. During this phase, engineers must use the existing designs to fabricate the actual aircraft. It specifically determines the design, location and quantity of elements such as rubs spars, sections and more. Furthermore, all aerodynamic, structural, control and performance aspects are achieved and tested in full in the previous preliminary design stage. The detail design stage, rather, is used to turn these designs into a working aircraft. It may also include flight simulations to test the design and ensure it functions as intended.

Aircraft Design Characteristics

Each aircraft handles somewhat differently because each resists or responds to control pressures in its own way. For example, a training aircraft is quick to respond to control applications, while a transport aircraft feels heavy on the controls and responds to control pressures more slowly. These features can be designed into an aircraft to facilitate the particular purpose of the aircraft by considering certain stability and maneuvering requirements.

Stability

Stability is the inherent quality of an aircraft to correct for conditions that may disturb its equilibrium and to return to or to continue on the original flight path. It is primarily an aircraft design characteristic. The flight paths and attitudes an aircraft flies are limited by the aerodynamic characteristics of the aircraft, its propulsion system, and its structural strength. These limitations indicate the maximum performance and maneuverability of the aircraft. If the aircraft is to provide maximum utility, it must be safely controllable to the full extent of these limits without exceeding the pilot's strength or requiring exceptional flying ability. If an aircraft is to fly straight and steady along any arbitrary flight path, the forces acting on it must be in static equilibrium. The reaction of anybody when its equilibrium is disturbed is referred to as stability. The two types of stability are static and dynamic.

Static Stability

Static stability refers to the initial tendency, or direction of movement, back to equilibrium. In aviation, it refers to the aircraft's initial response when disturbed from a given pitch, yaw, or bank.

- Positive static stability: The initial tendency of the aircraft to return to the original state of equilibrium after being disturbed.

- Neutral static stability: The initial tendency of the aircraft to remain in a new condition after its equilibrium has been disturbed.
- Negative static stability: The initial tendency of the aircraft to continue away from the original state of equilibrium after being disturbed.

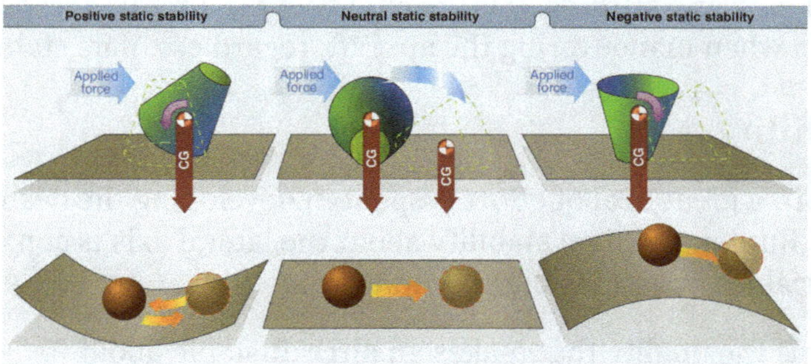

Types of static stability.

Dynamic Stability

Static stability has been defined as the initial tendency to return to equilibrium that the aircraft displays after being disturbed from its trimmed condition. Occasionally, the initial tendency is different or opposite from the overall tendency, so a distinction must be made between the two. Dynamic stability refers to the aircraft response over time when disturbed from a given pitch, yaw, or bank. This type of stability also has three subtypes:

- Positive dynamic stability: Over time, the motion of the displaced object decreases in amplitude and, because it is positive, the object displaced returns toward the equilibrium state.
- Neutral dynamic stability: Once displaced, the displaced object neither decreases nor increases in amplitude. A worn automobile shock absorber exhibits this tendency.
- Negative dynamic stability: Over time, the motion of the displaced object increases and becomes more divergent.

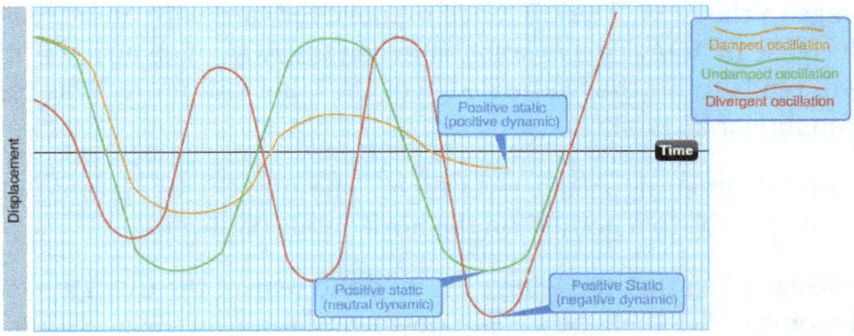

Damped versus undamped stability.

Stability in an aircraft affects two areas significantly:

- Maneuverability: The quality of an aircraft that permits it to be maneuvered easily and to withstand the stresses imposed by maneuvers. It is governed by the aircraft's weight, inertia, size and location of flight controls, structural strength, and power plant. It too is an aircraft design characteristic.

- Controllability: The capability of an aircraft to respond to the pilot's control, especially with regard to flight path and attitude. It is the quality of the aircraft's response to the pilot's control application when maneuvering the aircraft, regardless of its stability characteristics.

Longitudinal Stability

In designing an aircraft, a great deal of effort is spent in developing the desired degree of stability around all three axes. But longitudinal stability about the lateral axis is considered to be the most affected by certain variables in various flight conditions.

Longitudinal stability is the quality that makes an aircraft stable about its lateral axis. It involves the pitching motion as the aircraft's nose moves up and down in flight. A longitudinally unstable aircraft has a tendency to dive or climb progressively into a very steep dive or climb, or even a stall. Thus, an aircraft with longitudinal instability becomes difficult and sometimes dangerous to fly.

Static longitudinal stability, or instability in an aircraft, is dependent upon three factors:

- Location of the wing with respect to the CG.
- Location of the horizontal tail surfaces with respect to the CG.
- Area or size of the tail surfaces.

In analyzing stability, it should be recalled that a body free to rotate always turns about its CG. To obtain static longitudinal stability, the relation of the wing and tail moments must be such that, if the moments are initially balanced and the aircraft is suddenly nose up, the wing moments and tail moments change so that the sum of their forces provides an unbalanced but restoring moment which, in turn, brings the nose down again. Similarly, if the aircraft is nose down, the resulting change in moments brings the nose back up.

The Center of Lift (CL) in most asymmetrical airfoils has a tendency to change its fore and aft positions with a change in the AOA. The CL tends to move forward with an increase in AOA and to move aft with a decrease in AOA. This means that when the AOA of an airfoil is increased, the CL, by moving forward, tends to lift the leading edge of the wing still more. This tendency gives the wing an inherent quality of instability.

Figure shows an aircraft in straight-and-level flight. The line CG-CL-T represents the aircraft's longitudinal axis from the CG to a point T on the horizontal stabilizer.

Most aircraft are designed so that the wing's CL is to the rear of the CG. This makes the aircraft "nose heavy" and requires that there be a slight downward force on the horizontal stabilizer in order to balance the aircraft and keep the nose from continually pitching downward. Compensation for this nose heaviness is provided by setting the horizontal stabilizer at a slight negative AOA. The downward force thus produced holds the tail down, counterbalancing the "heavy" nose. It is as if the line CG-CL-T were a lever with an upward force at CL and two downward forces balancing each other, one a strong force at the CG point and the other, a much lesser force, at point T (downward air pressure on the stabilizer). To better visualize this physics principle: If an iron bar were suspended at point CL, with a heavy weight hanging on it at the CG, it would take downward pressure at point T to keep the "lever" in balance.

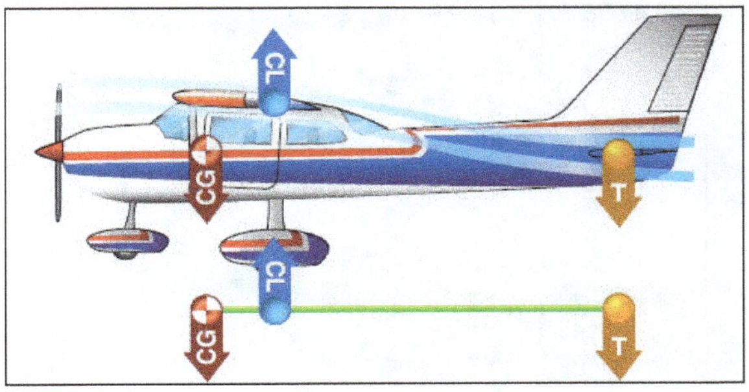

Longitudinal stability.

Even though the horizontal stabilizer may be level when the aircraft is in level flight, there is a downwash of air from the wings. This downwash strikes the top of the stabilizer and produces a downward pressure, which at a certain speed is just enough to balance the "lever." The faster the aircraft is flying, the greater this downwash and the greater the downward force on the horizontal stabilizer (except T-tails). In aircraft with fixed-position horizontal stabilizers, the aircraft manufacturer sets the stabilizer at an angle that provides the best stability (or balance) during flight at the design cruising speed and power setting.

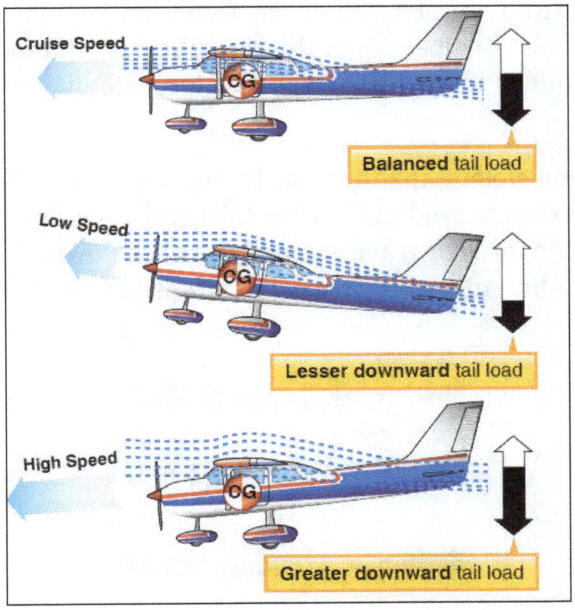

If the aircraft's speed decreases, the speed of the airflow over the wing is decreased. As a result of this decreased flow of air over the wing, the downwash is reduced, causing a lesser downward force on the horizontal stabilizer. In turn, the characteristic nose heaviness is accentuated, causing the aircraft's nose to pitch down more. This places the aircraft in a nose-low attitude, lessening the wing's AOA and drag and allowing the airspeed to increase. As the aircraft continues in the nose-low attitude and its speed increases, the downward force on the horizontal stabilizer is once again increased. Consequently, the tail is again pushed downward and the nose rises into a climbing attitude.

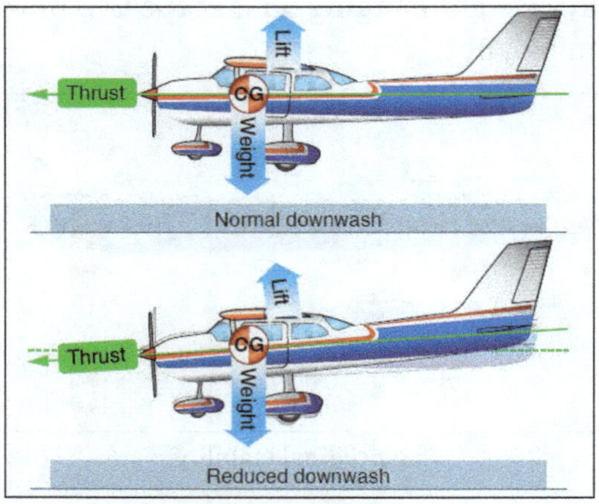

Reduced power allows pitch down.

As this climb continues, the airspeed again decreases, causing the downward force on the tail to decrease until the nose lowers once more. Because the aircraft is dynamically stable, the nose does not lower as far this time as it did before. The aircraft acquires enough speed in this more gradual dive to start it into another climb, but the climb is not as steep as the preceding one.

After several of these diminishing oscillations, in which the nose alternately rises and lowers, the aircraft finally settles down to a speed at which the downward force on the tail exactly counteracts the tendency of the aircraft to dive. When this condition is attained, the aircraft is once again in balanced flight and continues in stabilized flight as long as this attitude and airspeed are not changed.

A similar effect is noted upon closing the throttle. The downwash of the wings is reduced and the force at T in figure is not enough to hold the horizontal stabilizer down. It seems as if the force at T on the lever were allowing the force of gravity to pull the nose down. This is a desirable characteristic because the aircraft is inherently trying to regain airspeed and reestablish the proper balance.

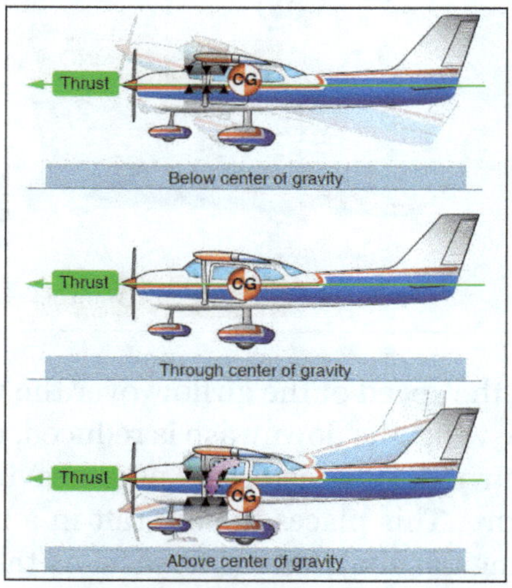

Thrust line affects longitudinal stability.

Power or thrust can also have a destabilizing effect in that an increase of power may tend to make the nose rise. The aircraft designer can offset this by establishing a "high thrust line" wherein the line of thrust passes above the CG. In this case, as power or thrust is increased a moment is produced to counteract the down load on the tail. On the other hand, a very "low thrust line" would tend to add to the nose-up effect of the horizontal tail surface. CG forward of the CL and with an aerodynamic tail-down force, the aircraft usually tries to return to a safe flying attitude.

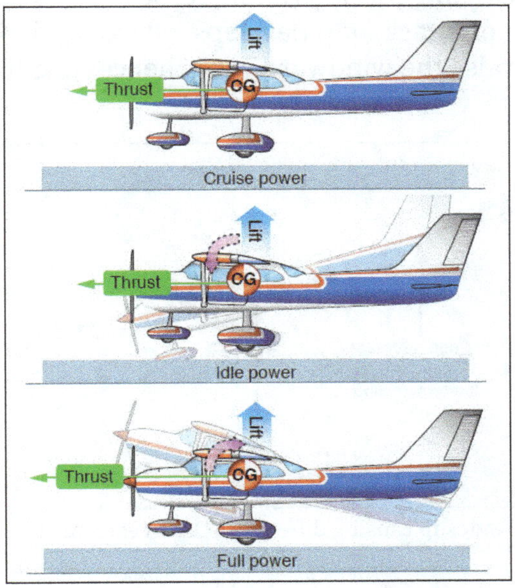

Power changes affect longitudinal stability.

The following is a simple demonstration of longitudinal stability. Trim the aircraft for "hands off" control in level flight. Then, momentarily give the controls a slight push to nose the aircraft down. If, within a brief period, the nose rises towards the original position, the aircraft is statically stable. Ordinarily, the nose passes the original position (that of level flight) and a series of slow pitching oscillations follows. If the oscillations gradually cease, the aircraft has positive stability; if they continue unevenly, the aircraft has neutral stability; if they increase, the aircraft is unstable.

Lateral Stability (Rolling)

Stability about the aircraft's longitudinal axis, which extends from the nose of the aircraft to its tail, is called lateral stability. Positive lateral stability helps to stabilize the lateral or "rolling effect" when one wing gets lower than the wing on the opposite side of the aircraft. There are four main design factors that make an aircraft laterally stable: dihedral, sweepback, keel effect, and weight distribution.

Dihedral is the upward angle of the wings from a horizontal (front/rear view) axis of the plane.

Dihedral

Some aircraft are designed so that the outer tips of the wings are higher than the wing roots. The upward angle thus formed by the wings is called dihedral. When a gust causes a roll, a sideslip will result. This sideslip causes the relative wind affecting the entire airplane to be from the direction of the slip. When the relative wind comes from the side, the wing slipping into the wind is subject to an increase in AOA and develops an increase in lift. The wing away from the wind is subject to a decrease in angle of attack, and develops a decrease in lift. The changes in lift effect a rolling moment tending to raise the windward wing, hence dihedral contributes to a stable roll due to sideslip.

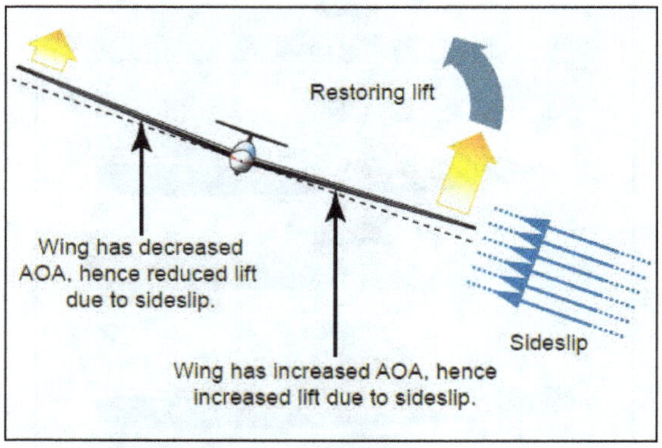

Sideslip causing different AOA on each blade.

Sweepback and Wing Location

Many aspects of an aircraft's configuration can affect its effective dihedral, but two major components are wing sweepback and the wing location with respect to the fuselage (such as a low wing or high wing). As a rough estimation, 10° of sweepback on a wing provides about 1° of effective dihedral, while a high wing configuration can provide about 5° of effective dihedral over a low wing configuration.

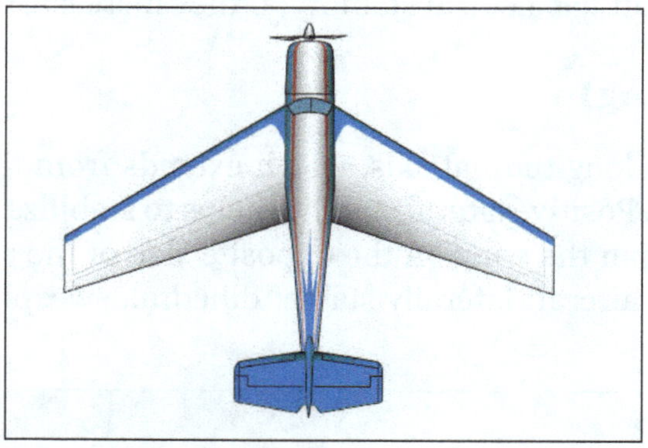

Sweepback wings.

A sweptback wing is one in which the leading edge slopes backward. When a disturbance causes an aircraft with sweepback to slip or drop a wing, the low wing presents its leading edge at an angle that is more perpendicular to the relative airflow. As a result, the low wing acquires more lift, rises, and the aircraft is restored to its original flight attitude.

Keel Effect and Weight Distribution

A high wing aircraft always has the tendency to turn the longitudinal axis of the aircraft into the relative wind, which is often referred to as the keel effect. These aircraft are laterally stable simply because the wings are attached in a high position on the fuselage, making the fuselage behave like a keel exerting a steadying influence on the aircraft laterally about the longitudinal axis. When a high-winged aircraft is disturbed and one wing dips, the fuselage weight acts like a pendulum returning the aircraft to the horizontal level.

Laterally stable aircraft are constructed so that the greater portion of the keel area is above the CG. Thus, when the aircraft slips to one side, the combination of the aircraft's weight and the pressure of the airflow against the upper portion of the keel area (both acting about the CG) tends to roll the aircraft back to wings-level flight.

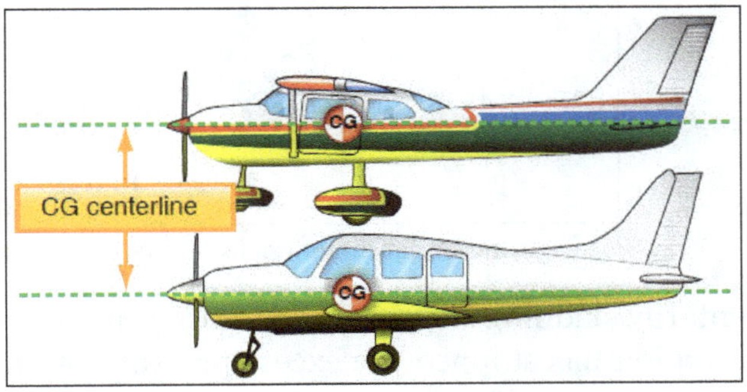

Keel area for lateral stability.

Directional Stability (Yawing)

Stability about the aircraft's vertical axis (the sideways moment) is called yawing or directional stability. Yawing or directional stability is the most easily achieved stability in aircraft design. The area of the vertical fin and the sides of the fuselage aft of the CG are the prime contributors that make the aircraft act like the well-known weather vane or arrow, pointing its nose into the relative wind.

In examining a weather vane, it can be seen that if exactly the same amount of surface were exposed to the wind in front of the pivot point as behind it, the forces fore and aft would be in balance and little or no directional movement would result. Consequently, it is necessary to have a greater surface aft of the pivot point than forward of it.

Similarly, the aircraft designer must ensure positive directional stability by making the side surface greater aft than ahead of the CG. To provide additional positive stability to that provided by the fuselage, a vertical fin is added. The fin acts similar to the feather on an arrow in maintaining straight flight. Like the weather vane and the arrow, the farther aft this fin is placed and the larger its size, the greater the aircraft's directional stability.

If an aircraft is flying in a straight line, and a sideward gust of air gives the aircraft a slight rotation about its vertical axis (i.e., the right), the motion is retarded and stopped by the fin because while the aircraft is rotating to the right, the air is striking the left side of the fin at an angle. This causes pressure on the left side of the fin, which resists the turning motion and slows down the aircraft's yaw. In doing so, it acts somewhat like the weather vane by turning the aircraft into the relative wind. The initial change in direction of the aircraft's flight path is generally slightly behind its change of heading. Therefore, after a slight yawing of the aircraft to the right, there is a brief moment when the aircraft is still moving along its original path, but its longitudinal axis is pointed slightly to the right.

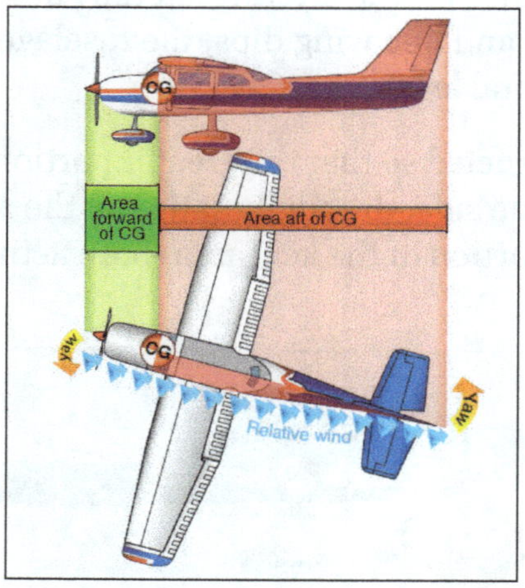

Fuselage and fin for directional stability.

The aircraft is then momentarily skidding sideways and, during that moment (since it is assumed that although the yawing motion has stopped the excess pressure on the left side of the fin still persists), there is necessarily a tendency for the aircraft to be turned partially back to the left. That is, there is a momentary restoring tendency caused by the fin.

This restoring tendency is relatively slow in developing and ceases when the aircraft stops skidding. When it ceases, the aircraft is flying in a direction slightly different from the original direction. In other words, it will not return of its own accord to the original heading; the pilot must reestablish the initial heading.

A minor improvement of directional stability may be obtained through sweepback. Sweepback is incorporated in the design of the wing primarily to delay the onset of compressibility during high-speed flight. In lighter and slower aircraft, sweepback aids in locating the center of pressure in the correct relationship with the CG. A longitudinally stable aircraft is built with the center of pressure aft of the CG.

Because of structural reasons, aircraft designers sometimes cannot attach the wings to the fuselage at the exact desired point. If they had to mount the wings too far forward, and at right angles to the fuselage, the center of pressure would not be far enough to the rear to result in the desired amount of longitudinal stability. By building sweepback into the wings, however, the designers can move the center of pressure toward the rear. The amount of sweepback and the position of the wings then place the center of pressure in the correct location.

When turbulence or rudder application causes the aircraft to yaw to one side, the opposite wing presents a longer leading edge perpendicular to the relative airflow. The airspeed of the forward wing increases and it acquires more drag than the back wing. The additional drag on the forward wing pulls the wing back, turning the aircraft back to its original path.

The contribution of the wing to static directional stability is usually small. The swept wing provides a stable contribution depending on the amount of sweepback, but the contribution is relatively small when compared with other components.

Free Directional Oscillations (Dutch Roll)

Dutch roll is a coupled lateral/directional oscillation that is usually dynamically stable but is unsafe in an aircraft because of the oscillatory nature. The damping of the oscillatory mode may be weak or strong depending on the properties of the particular aircraft.

If the aircraft has a right wing pushed down, the positive sideslip angle corrects the wing laterally before the nose is realigned with the relative wind. As the wing corrects the position, a lateral directional oscillation can occur resulting in the nose of the aircraft making a figure eight on the horizon as a result of two oscillations (roll and yaw), which, although of about the same magnitude, are out of phase with each other.

In most modern aircraft, except high-speed swept wing designs, these free directional oscillations usually die out automatically in very few cycles unless the air continues to be gusty or turbulent. Those aircraft with continuing Dutch roll tendencies are usually equipped with gyro-stabilized yaw dampers. Manufacturers try to reach a midpoint between too much and too little directional stability. Because it is more desirable for the aircraft to have "spiral instability" than Dutch roll tendencies, most aircraft are designed with that characteristic.

Spiral Instability

Spiral instability exists when the static directional stability of the aircraft is very strong as compared to the effect of its dihedral in maintaining lateral equilibrium. When the lateral equilibrium of the aircraft is disturbed by a gust of air and a sideslip is introduced, the strong directional stability tends to yaw the nose into the resultant relative wind while the comparatively weak dihedral lags in restoring the lateral balance. Due to this yaw, the wing on the outside of the turning moment travels forward faster than the inside wing and, as a consequence, its lift becomes greater. This produces an overbanking tendency which, if not corrected by the pilot, results in the bank angle becoming steeper and steeper. At the same time, the strong directional stability that yaws the aircraft into the relative wind is actually forcing the nose to a lower pitch attitude. A slow downward spiral begins which, if not counteracted by the pilot, gradually increases into a steep spiral dive. Usually the rate of divergence in the spiral motion is so gradual the pilot can control the tendency without any difficulty.

Many aircraft are affected to some degree by this characteristic, although they may be inherently stable in all other normal parameters. This tendency explains why an aircraft cannot be flown "hands off" indefinitely.

Materials used in Aircrafts

It is observed that there is currently a great concern among engineers in selecting materials for aircraft industry. This is because airplanes are a means of transport that require high levels of security, coupled with the performance maximization. Most of the aircraft structure is built with titanium, steel and aluminium. However there is a great tendency for further use of new light-weight materials based on polymers and carbon fiber. These combinations of plastics and carbon fiber are called composite materials, and have low weight and high strength. New alloys are also being developed to suit modern requirements. Even so-called composite aircraft are around 50% composite meaning the other 50% is still metallic.

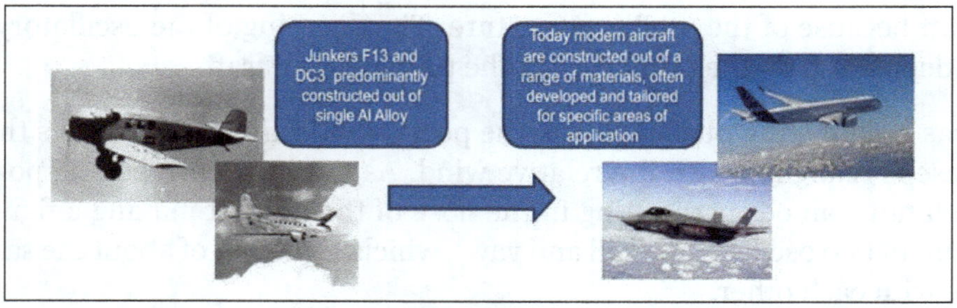

Evolution in use in the use of materials in the aerospace industry.

The issue of the use of new materials in the aerospace industry goes far beyond to reduce weight and improve performance, is also related to commercial and market factors. Currently we are looking for new materials that meet the following prerequisites: Harder materials, low cost, larger more complex structures, environmental considerations. The figure below illustrates how different materials are used in a Boing 787 and compares with the materials used in the Boeing 777 from the previous generation.

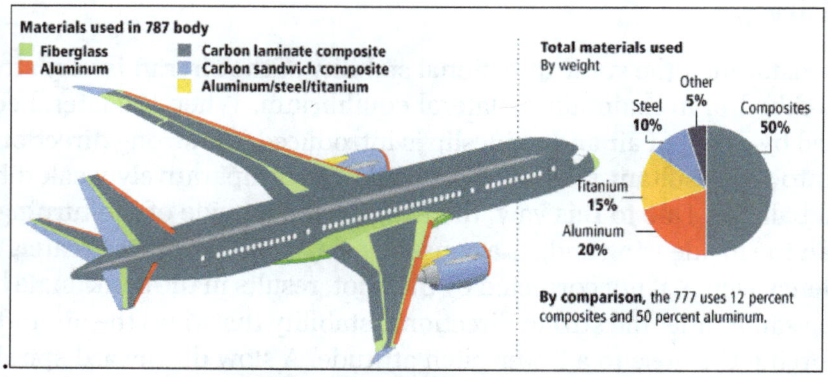

Use of different materials in a Boeing 787.

In addition to the use of new materials, also new techniques are being used in favour the aerospace industry. Today we have many more options of welding processes that guarantee a much higher quality to the final piece. The welding processes with more added technology that are used today are: Laser welding and friction stir welding. Further processes such as powder metallurgy are gaining ground in this field. We have also made substantial progress in the area of simulations, that way you can predict the behavior of a material without the need to test it in practice, saving time and ensuring quality.

New Metallic Materials

In the context of new materials we highlight the development of new alloys and the use of noble metals. The use of titanium is the one that has grown in recent years. Another important consideration is that despite we are moving forward the use of lighter alloys and high performance corrosion resistance of these materials has remained constant or even decreased, thus it is necessary to use strategies such as the use of coatings to prevent the rapid degradation of portions of the aircraft. The chart below shows how the use of alloys has in aerospace and the corrosion resistance of each league:

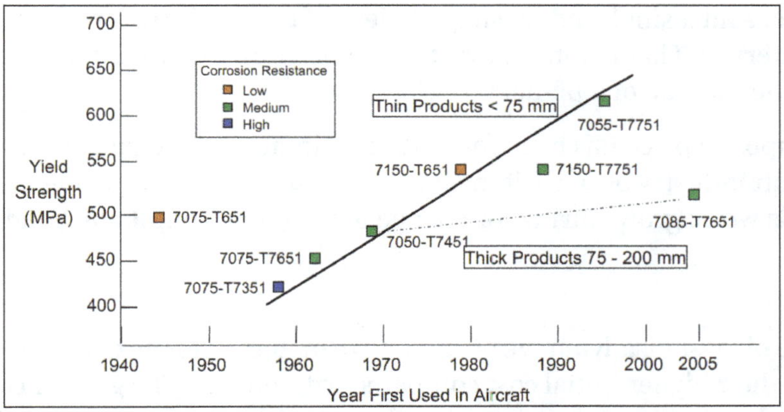

Graph relating yield strength, years of use of the alloy and corrosion resistance.

New Composite Materials

Historically, composite aircraft structures have been fabricated using long fibres having a length equal to the dimension of the component, but recent studies have combined the use of preforms and dried chopped fibres and resin transfer processes. Although the epoxy matrix is still widely used in the processing of composites, a new generation of modified thermosetting resin has been used. An example is the supersonic aircraft F22, which achieves high speeds; it uses 24% of its structure composite polymeric material, 39% titanium, 16% aluminium, steel 6% and 15% of other materials.

Advantages and Disadvantages of using Composite Materials in Airplanes

Composite materials—especially those made from glass fibers, carbon fibers, and Kevlar—are widely used in the aircraft industry. They are stronger and lighter than aluminum, the metal that's most commonly used in aircraft bodies.

Composite materials are so named because they consist of two or more materials. The composites used in planes consist of fibers suspended in a matrix of epoxy resin.

Composite materials have revolutionized aviation, but their use does present some engineering and maintenance challenges. Here is a roundup of the advantages and disadvantages of using composites in aircraft.

Advantages

- Weight reduction is the single greatest advantage of composite material usage. A lower-weight plane is more fuel-efficient because it requires less fuel to propel itself forward.
- Composites are also incredibly strong and as a result have a higher strength-to-weight ratio, also known as specific strength, than the metals used in making aircraft. In addition, they resist compression and don't easily break under tension.

- Composite materials aren't prone to corrosion due to harsh chemicals, and they're resistant to many highly reactive chemicals. They can also handle wide variations in temperature and exposure to severe weather.
- Another big advantage of composites is their design flexibility: They can be made into just about shape. And a single, oddly shaped piece of composite can replace many pieces made of other materials. That helpful characteristic cuts down on maintenance and so can reduce costs over the lifetime of a plane.
- Once a composite piece has been formed, it maintains its shape and size. That's important in the aircraft industry because it means the essential parts of a plane made out of composite materials won't grow, shrink, or change form as environmental conditions vary.

Disadvantages

- Perhaps the biggest disadvantage of composite materials for aircraft and component manufacturers is their higher initial cost compared with metals. The greater cost is largely due to the price of the fibers and the complicated process required making the finished materials.
- It can be difficult to tell when the interior structure of a composite aircraft piece has been damaged. That makes inspections difficult and more costly.
- One issue that comes up during inspections is delamination—when layers of composites separate. The biggest cause of delamination is an impact to the composite piece. Water can infiltrate a piece that has experienced delamination, and the problem will get worse as the water freezes and thaws.
- Standard aircraft composite materials do not conduct electricity and so, unlike aluminum, cannot prevent lightning from being directed to a plane's fuel tanks.
- Finally, the resin used in composite materials weakens at temperatures as low as 150 degrees Fahrenheit, and a temperature above 300 degrees might result in aircraft failure. When composite materials burn, they give off toxic fumes and micro-particles into the air, causing health risks. For all of those reasons, fire prevention is crucial on planes built with composites.

Parts of an Aircraft

Cockpit

The cockpit is the area where the pilots and crew sit to fly an airplane. In a small plane, the cockpit might be occupied by a single pilot.

Parts of the Cockpit

Cockpit

The most obvious features in a cockpit are the seats for the pilots and the instrumentation panel. However, you will also be able to find stowage areas (e.g. for documents, fire extinguishers, clothing), as well as the pedestal, side consoles, a third (or jump) seat and access to electric and electronic compartments.

Cockpit Enclosure

This is a typical layout for an aircraft cockpit. Yet, the design may change depending on manufacturers' preferences and type of aircraft.

Seats

In modern planes, the seats in the cockpit are designed ergonomically, for comfort during long flights. A third, or jump seat, may be present, which can be used by observers (e.g. doing LOSA) or invited people (e.g. other pilots).

Flight Deck

The flight deck comprises the main displays and controls for flying the aircraft, as well as their location in the cockpit environment. A typical airliner will have the main displays in the instrumentation panel, in front of the pilots, and the controls around him (central pedestal and side consoles) and above him (overhead panel). Other compartments are accessories and maintenance panel.

Control Column

A yoke is, perhaps, the prototypical flight control, positioned right in front of each pilot. It controls pitch (nose up/down via pull/push inputs) and roll (left/right bank via left/right turn inputs), and may control trim, as well. However, pitch and bank can also be controlled via a stick, as in some Airbus's models.

Rudder Pedals

They are positioned on the floor in front of the pilots, and act on the rudder, at the tail of the aircraft. They control jaw (right/left movement via push inputs on the right/left pedal) while flying, as well as steer the aircraft on the ground.

Instrument Panel

The main instrument panel holds the most important flight displays regarding both flight performance and aircraft status.

Pedestal

It normally contains the throttle and other engine controls, input keyboards for the navigational system, etc.

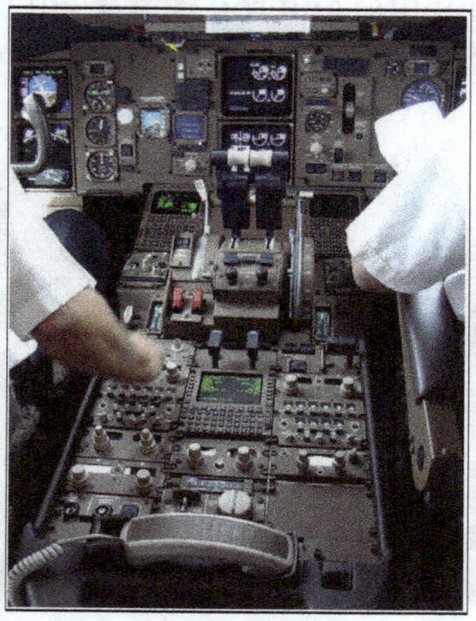

Side Consoles

They are used for placing the side stick, communication instruments, documentation, etc., depending on the type of airplane.

Overhead Panel

It contains miscellaneous controls and features not directly involved in flying. E.g. air conditioning, anti-ice controls, cabin pressurization controls, oxygen controls, etc.

Fuselage

The fuselage is an aircraft's main body section. It holds crew, passengers, and cargo. In single-engine aircraft it will usually contain an engine, as well, although in some amphibious aircraft the single engine is mounted on a pylon attached to the fuselage, which in turn is used as a floating hull. The fuselage also serves to position control and stabilization surfaces in specific relationships to lifting surfaces, which is required for aircraft stability and maneuverability.

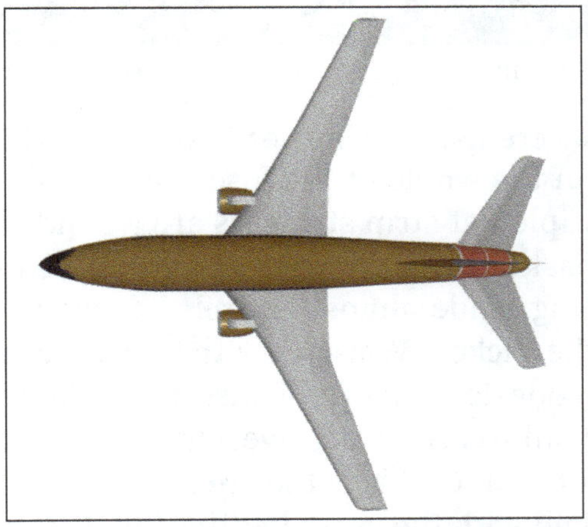

Fuselage of a Boeing 737 shown in brown.

Types of Structures

Truss Structure

Piper PA-18 welded tube truss fuselage structure.

This type of structure is still in use in many lightweight aircraft using welded steel tube trusses. A box truss fuselage structure can also be built out of wood—often covered with plywood. Simple box structures may be rounded by the addition of supported lightweight stringers, allowing the fabric covering to form a more aerodynamic shape, or one more pleasing to the eye.

Geodesic Construction

Geodesic airframe fuselage structure is exposed by battle damage.

Geodesic structural elements were used by Barnes Wallis for British Vickers between the wars and into World War II to form the whole of the fuselage, including its aerodynamic shape. In this type of construction multiple flat strip stringers are wound about the formers in opposite spiral directions, forming a basket-like appearance. This proved to be light, strong, and rigid and had the advantage of being made almost entirely of wood. A similar construction using aluminum alloy was used in the Vickers Warwick with less material than would be required for other structural types. The geodesic structure is also redundant and so can survive localized damage without catastrophic failure. A fabric covering over the structure completed the aerodynamic shell. The logical evolution of this is the creation of fuselages using molded plywood, in which multiple sheets are laid with the grain in differing directions to give the monocoque.

Monocoque Shell

In this method, the exterior surface of the fuselage is also the primary structure. A typical early form of this was built using molded plywood, where the layers of plywood are formed over a "plug" or within a mold. A later form of this structure uses fiberglass cloth impregnated with polyester or epoxy resin, instead of plywood, as the skin. A simple form of this used in some amateur-built aircraft uses rigid expanded foam plastic as the core, with a fiberglass covering, eliminating the necessity of fabricating molds, but requiring more effort in finishing. An example of a larger molded plywood aircraft is the de Havilland Mosquito fighter/light bomber of World War II. No plywood-skin fuselage is truly monocoque, since stiffening elements are incorporated into the structure to carry concentrated loads that would otherwise buckle the thin skin. The use of molded fiberglass using negative ("female") molds (which give a nearly finished product) is prevalent in the series production of many modern sailplanes. The use of molded composites for fuselage structures is being extended to large passenger aircraft such as the Boeing 787 *Dreamliner* (using pressure-molding on female molds).

The Van's Aircraft RV-7 of semi-monocoque construction.

Semi-monocoque

Sectioned fuselage showing frames, stringers and skin all made of aluminium.

This is the preferred method of constructing an all-aluminum fuselage. First, a series of frames in the shape of the fuselage cross sections are held in position on a rigid fixture. These frames are then joined with lightweight longitudinal elements called stringers. These are in turn covered with a skin of sheet aluminum, attached by riveting or by bonding with special adhesives. The fixture is then disassembled and removed from the completed fuselage shell, which is then fitted out with wiring, controls, and interior equipment such as seats and luggage bins. Most modern large aircraft are built using this technique, but use several large sections constructed in this fashion which are then joined with fasteners to form the complete fuselage. As the accuracy of the final product is determined largely by the costly fixture, this form is suitable for series production, where a large number of identical aircraft are to be produced. Early examples of this type include the Douglas Aircraft DC-2 and DC-3 civil aircraft and the Boeing B-17 Flying Fortress. Most metal light aircraft are constructed using this process.

Both monocoque and semi-monocoque are referred to as "stressed skin" structures as all or a portion of the external load (i.e. from wings and empennage, and from discrete masses such as the engine) is taken by the surface covering. In addition, the entire load from internal pressurization is carried (as *skin tension*) by the external skin.

The proportioning of loads between the components is a design choice dictated largely by the dimensions, strength, and elasticity of the components available for construction and whether or not a design is intended to be "self-jigging", not requiring a complete fixture for alignment.

Materials

Inside view of the wooden, fabric-covered fuselage of a Fisher FP-202.

Early aircraft were constructed of wood frames covered in fabric. As monoplanes became popular, metal frames improved the strength, which eventually led to all-metal-structure aircraft, with metal covering for all its exterior surfaces - this was first pioneered in the second half of 1915. Some modern aircraft are constructed with composite materials for major control surfaces, wings, or the entire fuselage such as the Boeing 787. On the 787, it makes possible higher pressurization levels and larger windows for passenger comfort as well as lower weight to reduce operating costs. The Boeing 787 weighs 1500 lb less than if it were an all-aluminum assembly.

Windows

Cockpit windshields on the Airbus A320 must withstand bird strikes up to 350 kt and are made of chemically strengthened glass. They are usually composed of three layers or plies, of glass or plastic : the inner two are 8 mm (0.3 in.) thick each and are structural, while the outer ply, about 3 mm thick, is a barrier against foreign object damage and abrasion, with often a hydrophobic coating. It must prevent fogging inside the cabin and de-ice from −50 °C (−58 °F), this was previously done with thin wires similar to a rear car window, but now is a transparent, nanometers thick coating of indium tin oxide siting between plies, electrically conductive and thus transmitting heat. Curved glass improves aerodynamics but sight criteria also needs larger panes. A cockpit windshield is composed of 4–6 panels, 35 kg (77 lb.) each on an Airbus A320. In its lifetime, an average aircraft goes through three or four windshields, and the market is shared evenly between OEM and higher margins aftermarket.

Cabin windows, made from much lighter than glass stretched acrylic glass, and consists of multiple panes: an outer one built to support four times the maximum cabin pressure, an inner one for redundancy and a scratch pane near the passenger. Acrylic is susceptible to crazing: a network of fine cracks appears but can be polished to restore optical transparency, removal and polishing typically undergo every 2–3 years for uncoated windows.

Wing Integration

"Flying wing" aircraft, such as the Northrop YB-49 Flying Wing and the Northrop B-2 Spirit bomber have no separate fuselage; instead what would be the fuselage is a thickened portion of the wing structure.

Conversely there have been a small number of aircraft designs which have no separate wing, but use the fuselage to generate lift. Examples include National Aeronautics and Space Administration's experimental lifting body designs and the Vought XF5U-1 Flying Flapjack.

A blended wing body can be considered a mixture of the above. It carries the useful load in a fuselage producing lift. A modern example is Boeing X-48. One of the earliest aircraft using this design approach is Burnelli CBY-3, which fuselage was airfoil shaped to produce lift.

Wing

Wing is an important part because it works to help in balancing and improving the aircraft's stability when flying. This is the part that allows the plane to go up. There are two wings that are joined by a fuselage. The shape of the wings is designed that way in order to help it fly. Aircraft wings are airfoils that create lift when moved rapidly through the air. Aircraft designers have created a variety of wings with different aerodynamic properties. Attached to the body of an aircraft at different angles, these wings come in different shapes.

Types of Aircraft Wings

Rectangular Wing

The rectangular wing is the simplest to manufacture. It is a non-tapered, straight wing that is mostly used in small aircrafts. This wing extends out from the aircraft's fuselage at right angles (approximately).

Piper pa-38.

A good example of an aircraft where a rectangular wing was used is the Piper PA 38. One major disadvantage of a rectangular wing is that it isn't aerodynamically efficient.

Elliptical Wing

Supermarine Spitfire.

The elliptical wing is aerodynamically most efficient because elliptical span wise lift distribution induces the lowest possible drag. However, the manufacturability of this aircraft wing is poor. One of the most famous aircrafts in which elliptical wing was used is the Supermarine Spitfire that ruled the skies during the Battle of Britain.

The elliptical wing wasn't originally designed to minimize drag induction, but rather it was made to house the landing gear along with ammunition and guns inside a wing. So, the wing had to be thin.

The ellipse was the shape that allowed for the thinnest possible wing, giving room inside to hold the necessary things. In aircrafts like the Seversky P-35, we can see a semi-elliptical wing that has a trailing or leading edge elliptical.

Tapered Wing

The tapered wing was designed by modifying the rectangular wing. The chord of the wing is varied across the span for approximate elliptical lift distribution.

North American Aviation P-51 Mustang.

While it isn't as efficient as the standard elliptical wing, it does offer a compromise between efficiency and manufacturability. The P-51 Mustang, which was utilized by the USAAF to fight against the Luftwaffe, used the tapered wing.

Delta Wing

This low aspect ratio wing is used in supersonic aircrafts. The main advantage of a delta wing is that it is efficient in all regimes (supersonic, subsonic, and transonic). Moreover, this type of wing offers a large area for the shape thereby improving maneuverability and reducing wing loading.

Dassault Mirage 2000.

The delta wing doesn't just offer efficient flight experience but is also strong structurally and provides large volume for fuel storage. This wing is also simple to manufacture and maintain.

However, like any other type of aircraft wing, delta wing also has some disadvantages. The main disadvantages of this aircraft wing include:

- Due to their low aspect ratio, delta wings induce high drag.
- At low speed – during landing and takeoff –, these wings have a high angle of attack mainly because, at such low speeds, vortices generate the lift. High stall angles of the delta wings compensate for this.

The Dassault Mirage 2000 is a prime example of an aircraft which uses tailless delta wings. Some aircrafts use tailed delta wings and one of the most famous of those aircrafts is the Russian MiG-21.

Eurofighter Typhoon T1.

Another variant of delta wing is the cropped delta and it is seen in the Eurofighter Typhoon T1. The tips of this variant are cut off for reducing drag at low speeds.

Another variant of delta wings which is popularly used in combat aircrafts is the double delta. Leading edge angle of the double data isn't constant but has two values.

Trapezoidal Wing

Lockheed Martin F 22A Raptor.

The trapezoidal wing design offers outstanding flight performance. The trailing edge of the wing sweeps forward while the leading edge sweeps back. This type of wing is commonly used in U.S. combat aircrafts.

This configuration offers highly efficient supersonic flights and has good stealth characteristics. The only problem is that the wing loading is high which results in reduced maneuverability. The trapezoidal wing is used in the famous F-22 Raptor jet.

Ogive Wing

The ogive wing design is used in very high-speed aircrafts. The complex mathematical shape of this aircraft wing is derived to minimize drag at supersonic speeds. Ogive wings offer excellent performance at supersonic speeds with minimal drag.

Aerospatiale BAC Concorde.

The main disadvantage of these types of aircraft wings is that they are very complex and manufacturing them is difficult. Moreover, their subsonic performance isn't satisfactory in comparison. The retired Aerospatiale-BAC Concorde used ogive wings.

Swept Back Wings

Boeing 787-9 Dreamliner – Steve Lynes.

The aircraft wings whose leading edges are swept back are called swept back wings. Swept back wings reduce drag when an aircraft is flying at transonic speeds. The majority of high-speed commercial aircrafts use swept back wings. Boeing 787 Dreamliner is one example out of many that uses swept back wings.

Forward-swept Wings

The aircraft wings whose leading edges are swept forward are called swept forward wings. One disadvantage of this type of configuration is that because of the flow characteristics of the wings, the outboard wings stall before the flaps. This can cause controllability issues. Swept forward wings were therefore only used in very few aircraft, like the Grumman X-29 Switch Blade.

The main issue that made this type of wing configuration unsuitable was that it produced wing twisting when it bent under load, putting greater stress on wing roots. The Sukoi Su-47 Berkut is one of the very few aircraft that used this wing.

Sukhoi SU-47 Berkut.

Variable Sweep Wings

Swept wings are mostly suitable for high speeds, like supersonic and transonic, while unswept wings work best for low speeds i.e. subsonic. Variable sweep wings were designed to optimize flight experience over a range of speeds.

General Dynamics F-111 Aardvark.

The mechanical complexity is the main issue with this aircraft wing. General Dynamics F-11 Aardvark is the first aircraft that used variable sweep wings.

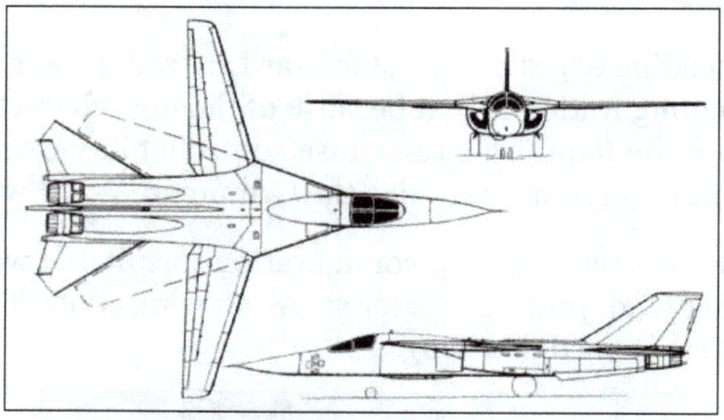

Aerofoil Section

An airfoil is the cross-sectional shape of a wing, blade (of a propeller, rotor, or turbine), or sail.

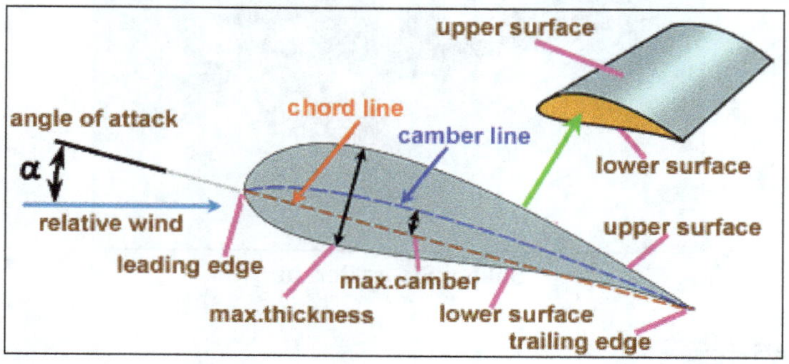

Different Wing Profiles

There are very many profiles which are classified according to their shapes by family.

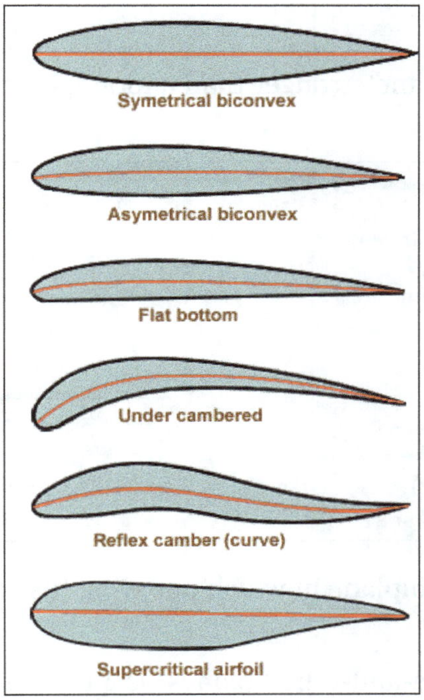

Number and Position of Main Planes

Fixed-wing aircraft can have different numbers of wings and the wing may be mounted at various positions relative to the fuselage.

Monoplane

- Low wing: Mounted near or below the bottom of the fuselage.
- Mid wing: Mounted approximately halfway up the fuselage.
- High wing: Mounted on the upper fuselage.
- Parasol wing: Raised clear above the top of the fuselage.

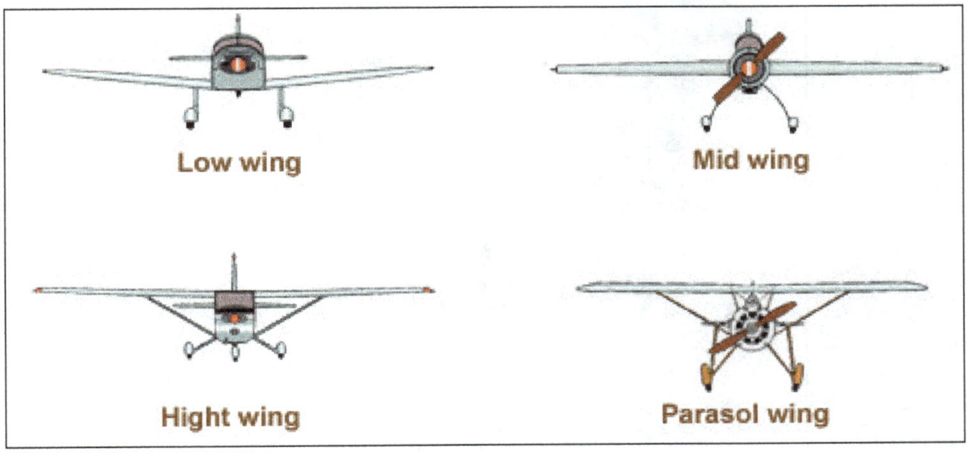

Multi-plane

The biplane is inherently lighter and stronger than a monoplane and was the most common configuration until the 1930s.

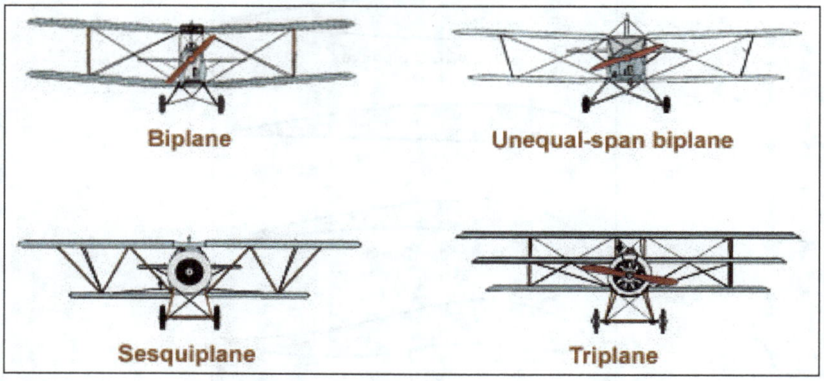

- Unequal-span biplane: a biplane in which one wing (usually the lower) is shorter than the other.
- Sesquiplane: literally "one-and-a-half planes" is a type of biplane in which the lower wing is significantly smaller than the upper wing, either in span or chord or both.
- Triplane: three planes stacked one above another.

Aspect Ratio

The aspect ratio is the span divided by the mean or average chord. It is a measure of how long and slender the wing appears.

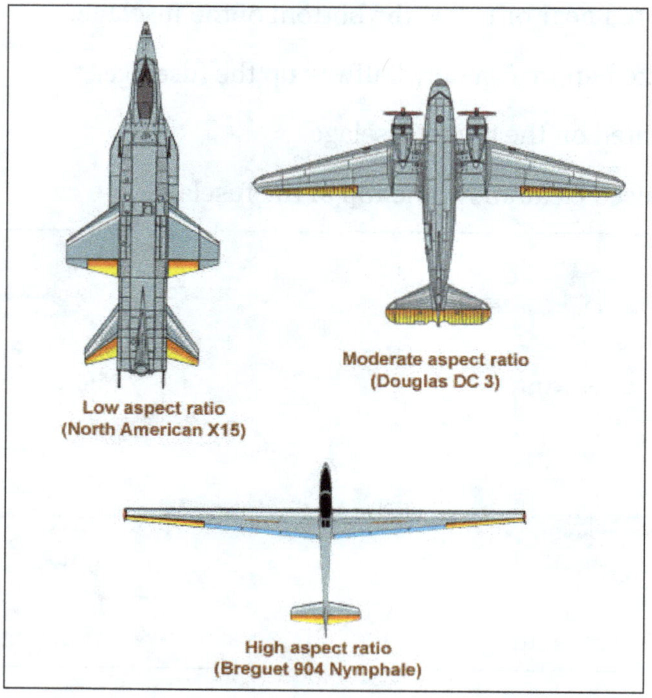

- Low aspect ratio: short and stubby wing. More efficient structurally and higher instantaneous roll rate. They tend to be used by fighter aircraft, such, and by very high-speed aircraft.
- Moderate aspect ratio: general-purpose wing, very widely used.
- High aspect ratio: long and slender wing. More efficient aerodynamically, having less induced drag. They tend to be used by high-altitude subsonic aircraft such as airliners.

Different Wings Shapes

Wing design can vary to provide certain desirable flight characteristics. Control at various operating speeds, the amount of lift generated, balance, and stability all change as the shape of the wing is altered. Both the leading edge and the trailing edge of the wing may be straight or curved, or one edge may be straight and the other curved. One or both edges may be tapered so that the wing is narrower at the tip than at the root where it joins the fuselage. The wing tip may be square, rounded, or even pointed.

Chord Variation Along Span

- Constant chord: parallel leading & trailing edges. Simplest to make, and common where low cost is important. Constant chord with Tapered Outer: his is midway between the rectangular and the tapered wing. Common variant seen for example on many Cessna types.
- Tapered: wing narrows towards the tip. Structurally and aerodynamically more efficient than a constant chord wing, and easier to make than the elliptical type.
    - Trapezoidal: a tapered wing with straight leading and trailing edges: may be unswept or swept. The straight tapered wing is one of the most common wing planforms.
- Elliptical: aerodynamically, the elliptical planform is the most efficient as elliptical spanwise lift distribution has the lowest possible induced drag. The most important disadvantage is that its manufacturability is poor.

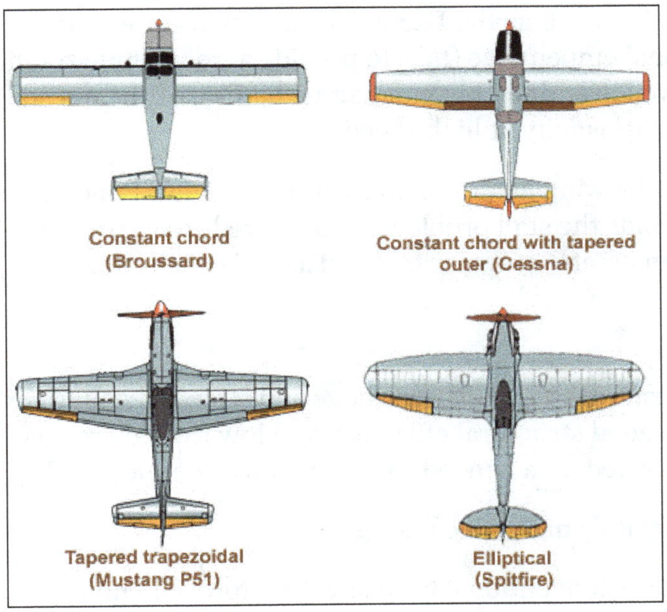

Wing Sweep

Wings may be swept back, or occasionally forwards, for a variety of reasons. A small degree of sweep is sometimes used to adjust the centre of lift when the wing cannot be attached in the ideal position for some reason, such as a pilot's visibility from the cockpit.

- Straight: extends at right angles to the line of flight. The most structurally-efficient wing, it has been common for low-speed designs since the very first days.
- Swept back (aka "swept wing"): The wing sweeps rearwards from the root to the tip. In early tailless examples, such as the Dunne aircraft, this allowed the outer wing section to act like a conventional empennage (tail) to provide aerodynamic stability. At transonic speeds swept wings have lower drag, but can handle badly in or near a stall and require high stiffness to avoid aeroelasticity at high speeds.
- Forward swept: the wing angles forward from the root. Benefits are similar to backwards sweep; also it avoids the stall problems and has reduced tip losses allowing a smaller wing, but requires even greater stiffness to avoid aeroelastic flutter.

Wing Delta

Delta: triangular planform with swept leading edge and straight trailing edge. Offers the advantages of a swept wing, with good structural efficiency and low frontal area. Disadvantages are the low wing loading and high wetted area needed to obtain aerodynamic stability.

- Tailless delta: a classic high-speed design.
- Tailed delta: adds a conventional tailplane, to improve handling.

- Cropped delta: wing tips are cut off. This helps avoid tip drag at high angles of attack.
- Compound delta or double delta: inner section has a (usually) steeper leading edge sweep. This improves the lift at high angles of attack and delays or prevents stalling.
- Ogival delta: a double-curve encompassing the leading edges and tip of a cropped compound delta.

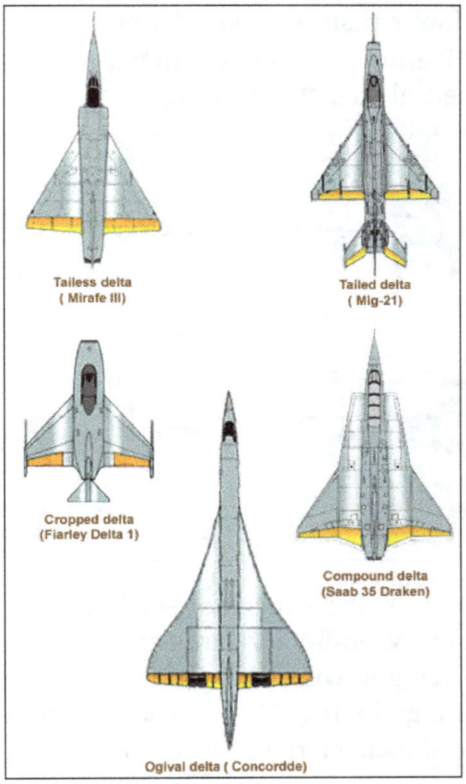

- Trapezoidal wing: This high-performance wing is characterized by low elongation with a positive-boom leading edge while the trailing edge has a negative boom.
- Flying wing: the aircraft has no distinct fuselage or horizontal tail (although fins and pods, blisters, etc. may be present.

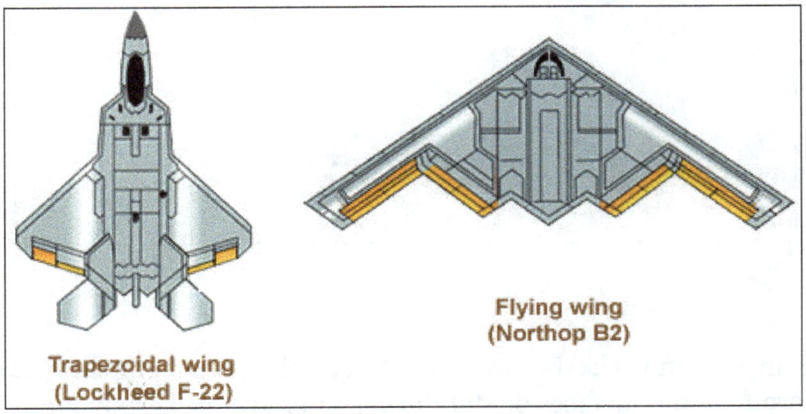

Dihedral and Anhedral

Angling the wings up or down spanwise from root to tip can help to resolve various design issues, such as stability and control in flight.

- Dihedral: The tips are higher than the root as giving a shallow 'V' shape when seen from the front. Adds lateral stability.
- Anhedral: The tips are lower than the root, the opposite of dihedral. Used to reduce stability where some other feature results in too much stability. Some biplanes have different degrees of dihedral/anhedral on different wings.

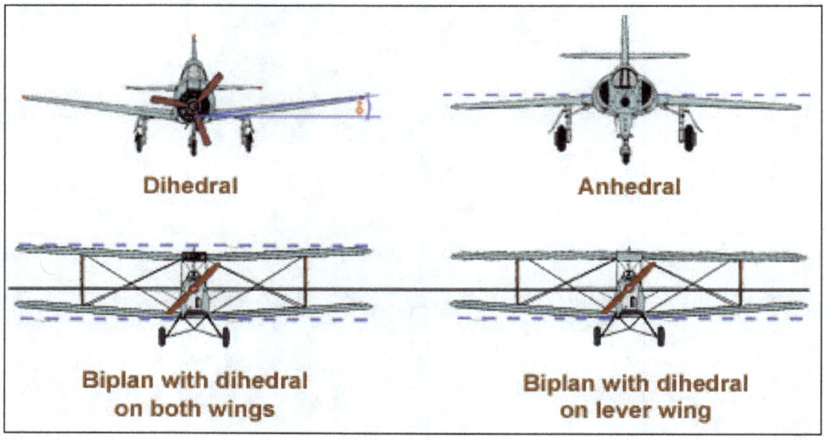

- Gull wing: Sharp dihedral on the wing root section, little or none on the main section. Sometimes used to improve visibility forwards and upwards and may be used as the upper wing on a biplane. Inverted gull wing, anhedral on the root section, dihedral on the main section. The opposite of a gull wing. May be used to reduce the length of wing-mounted undercarriage legs while allowing a raised fuselage.
- Cranked tip: Tip section dihedral differs from the main wing. The tips may have upwards dihedral or downwards anhedral.

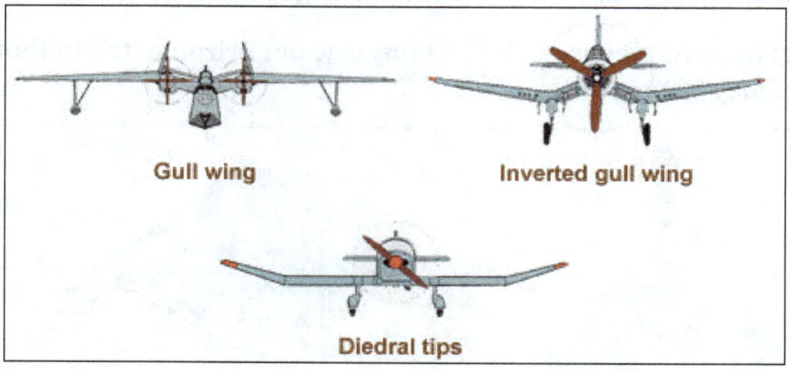

Empennage

The empennage of an aircraft is also known as the tail section. Most empennage designs consist of a tail cone, fixed aerodynamic surfaces or stabilizers, and movable aerodynamic surfaces.

The tail cone serves to close and streamline the aft end of most fuselages. The cone is made up of structural members like those of the fuselage; however, cones are usually of lighter construction since they receive less stress than the fuselage.

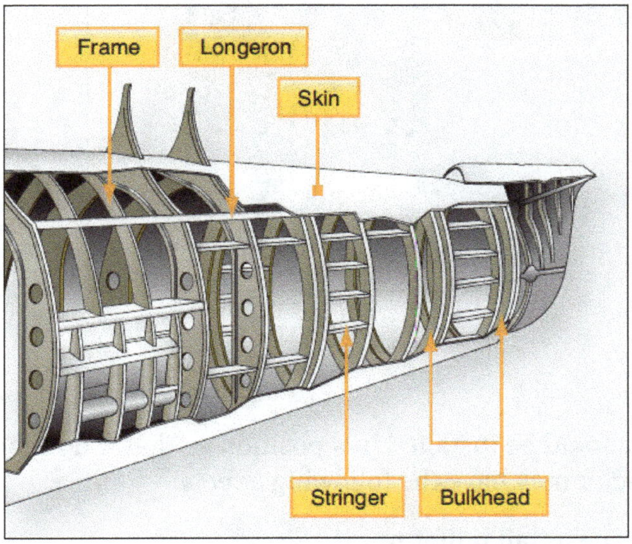

The fuselage terminates at the tail cone with similar but more lightweight construction.

The other components of the typical empennage are of heavier construction than the tail cone. These members include fixed surfaces that help stabilize the aircraft and movable surfaces that help to direct an aircraft during flight. The fixed surfaces are the horizontal stabilizer and vertical stabilizer. The movable surfaces are usually a rudder located at the aft edge of the vertical stabilizer and an elevator located at the aft edge the horizontal stabilizer.

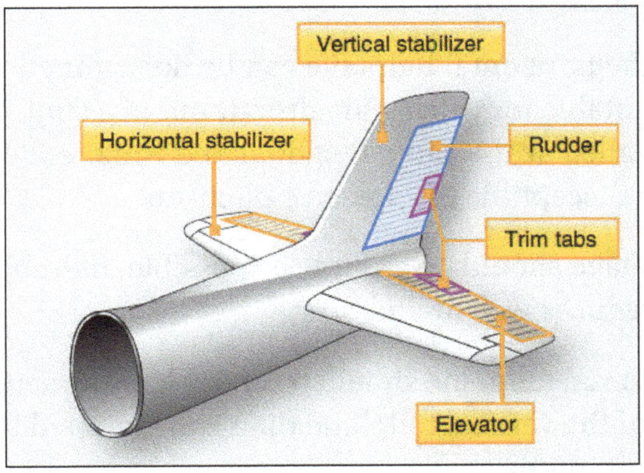

Components of a typical empennage.

The structure of the stabilizers is very similar to that which is used in wing construction. Figure shows a typical vertical stabilizer. Notice the use of spars, ribs, stringers, and skin like those found in a wing. They perform the same functions shaping and supporting the stabilizer and transferring stresses. Bending, torsion, and shear created by air loads in flight pass from one structural member to another. Each member absorbs some of the stress and passes the remainder on to the others. Ultimately, the spar transmits any overloads to the fuselage. A horizontal stabilizer is built the same way.

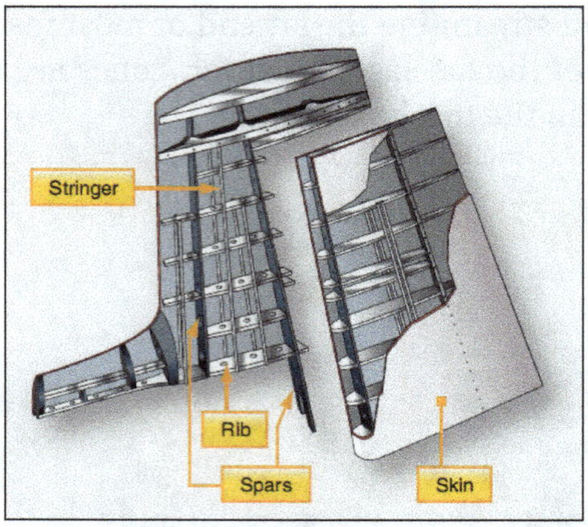

Vertical stabilizer.

Design Rules

The horizontal tailplane should be installed in a position so that it does not lie in the slip- stream. If this rule is not observed, it may have the following effects:

- Structure fatigue due to tail buffeting.
- Increased noise in the cabin due to tail buffeting.
- Considerable trim changes with differing choice of engine performance.

In some small single-engine aircraft the empennage is deliberately placed in the slipstream. Then one benefits from an increased efficiency of the tail assembly during take-off and landing, but may have to accept the disadvantages.

The detailed placement of the horizontal tailplane can be determined from figure: low-lying horizontal tailplanes are most suitable for getting an aircraft out of a stall. With sub- sonic aircraft the empennage can also be installed at the same height as the wing. A T-tail may only be used if the wing is uncritical and is not susceptible to excessive pitch-up.

The lever arm of the empennage should be as large as possible, thereby making it possible to keep the tail areas small, which reduces weight and drag.

The aspect ratio of the horizontal tailplane should be about half the aspect ratio of the wing. T-tails have a smaller aspect ratio of the vertical tailplane than conventional tails. This allows weight disadvantages to be kept to a minimum.

Tails with a taper ratio of $\lambda = 1$ are built in some cases as rectangular tail especially for general aviation aircraft. Rectangular tails reduce production costs.

The critical Mach number of the empennage $M_{crit\ H}$, und $M_{crit\ V}$, should be $\Delta M = 0\ 05$ higher than the critical Mach number of the wing $M_{crit\ W}$. Through this measure the efficiency of the tail assembly should also be guaranteed at high speed. Relative thickness, drag divergence Mach number, sweep, and the lift coefficient of the empennage must be chosen so as to ensure that a $\Delta M = 0\ 05$ can be achieved.

$$t/c = f\left(M_{DD}, \varphi_{25}, C_1, airfoil\right)$$

These parameters can be chosen to approximately suit each other if the drag divergence Mach number M_{DD} of the tail is $\Delta M = 0\ 05$ higher than for the wing.

The sweep of the horizontal tailplane should be approximately 5° larger than the sweep of the wing. Thus a higher critical Mach number of the horizontal tailplane can be achieved and a loss of efficiency due to shock waves is avoided. In addition, the lift gradient of the horizontal tailplane can be less than the lift gradient of the wing due to the increased sweep, so that the horizontal tailplane only reaches the stall state at larger angles of attack than the wing.

The sweep angle of the vertical tailplane is 35° to 55° for aircraft with "high airspeeds" (flight with compressibility effects). The sweep angle of the vertical tailplane for aircraft with "low airspeeds" (flight without compressibility effects) should be less than 20°. A large sweep angle increases the lever arm and the angle where the vertical tailplane goes into stall, but reduces the maximum lift coefficient.

The horizontal tailplane should have a relative thickness that is approximately 10 % less than the relative thickness in the outer wing. Thus, a higher critical Mach number of the horizontal tailplane can be achieved and a loss of efficiency due to shock waves is prevented.

Symmetrical airfoils are chosen exclusively for vertical tailplanes. Symmetrical or virtually symmetrical airfoils with 9% to 12% relative thickness are chosen for horizontal tailplanes. For example, NACA 0009 or NACA 0012 (Abbott 1959) can be chosen. Asymmetrical horizontal tailplane airfoils are installed "upside-down" because the horizontal tailplane has to create negative lift.

If the left and right elevators are to be connected, the sweep and the taper ratio must be selected so as to ensure that a hinge line is produced perpendicular to the aircraft's plane of symmetry. Reasons for connecting the elevators may be:

- To reduce the elevators' tendency to flutter.
- To facilitate joint actuation of the elevators.

The dihedral angle can be chosen so that the empennage is positioned outside the engine slipstream. Dihedral of the horizontal tail is not used to modify roll stability as this is much more influenced by the wing.

If the horizontal tailplane is fixed, an incidence angle of around 2° to 3° downwards should be chosen to create negative lift. A more flexible alternative is a movable, i.e. trimmable horizontal stabilizer, THS, which facilitates a larger center-of-gravity range.

The horizontal tailplane can be designed as an all moving tail. An all moving tail only consists of one surface with an adjustable incidence angle. The all moving tail is more effective – especially at high Mach numbers – but also heavier than a fixed empennage with control surface. In the case of large aircraft high output may be required to move the all moving tail in flight with the necessary actuating speed. A compromise is the trimmable horizontal stabilizer mentioned above: the horizontal stabilizer is used to trim and is only adjusted gradually (with a low actuating power); the elevator is deflected correspondingly quicker for maneuvering. The trimmable horizontal stabilizer is the standard solution for transport aircraft. Lifting canard or tandem wing is designed like wing.

Tables contain parameters that can be referred to as guides for empennage design. Conventional aspect ratios A and taper ratios λ from empennages on transport category airplanes.

Type	Horizontal Tailplane		Vertical Tailplane	
	A	λ	A	λ
Conventional Tail T-Tail	3.00 ... 5.00 as Conventional Tail	0.3 ... 0.6 as Conventional Tail	1.3 ... 2.0 0.7 ... 1.2	0.3 ... 0.6 0.6 ... 1.0

Table: Conventional design parameters for horizontal tails (Roskam II).

Type	Dihedral Angle v [°]	Incidence Angle ih [°]	Aspect Ratio Ah [-]	Sweep Angle φ [°]	Taper Ratio λh [-]
Business Jets	−4 ... 9	−3.5 fixed	3.2 ... 6.3	0 ... 35	0.32 ... 0.57
Transport Jets	0 ... 11	variable	3.4 ... 6.1	18 ... 37	0.27 ... 0.62
Fighters	−23 ... 5	0 fixed or variable	2.3 ... 5.8	0 ... 55	0.16 ... 1.00
Supersonic					
Civil Transport	-15 ... 0	0 fixed or variable	1.8 ... 2.6	32 ... 60	0.14 ... 0.39

Table: Conventional design parameters for vertical tails (Roskam II).

Type	Dihedral Angle v [°]	Incidence Angle ih [°]	Aspect Ratio Ah [-]	Sweep Angle φ [°]	Taper Tratio λh [-]
Business Jets	90	0	0.8 ... 1.6	28 ... 55	0.30 ... 0.74
Transport Jets	90	0	0.7 ... 2.0	33 ... 53	0.26 ... 0.73
Fighters	75 ... 90	0	0.4 ... 2.0	9 ... 60	0.19 ... 0.57
Supersonic Cruise					
Airplanes	75 ... 90	0	0.5 ... 1.8	37 ... 65	0.20 ... 0.43

Design According to Tail Volume

The area of the horizontal tailplane S_H or the vertical tailplane S_V multiplied by the lever arm l_H or l_V is called tail volume. The tail volume coefficient is defined for the horizontal tailplane as:

$$C_H = \frac{S_H . l_H}{S_W . c_{MAC}}$$

and for the vertical tailplane as:

$$C_V = \frac{S_V . l_V}{S_W . b}$$

- l_H the lever arm of the horizontal tailplane is the distance between the aerodynamic centers of wing and horizontal tailplane.
- l_V the lever arm of the vertical tailplane is the distance between the aerodynamic centers of wing and vertical tailplane.

As a good approximation the 25 % - point on the mean aerodynamic chord can also be referred to instead of the distances between the aerodynamic centers.

Table: Conventional tail volume coefficients of horizontal and vertical tails.

Type	Horizontal ch	Vertical c_v
General aviation - twin engine	0.80	0.07
Transport jets	1.00	0.08
Jet - trainer	0.70	0.06
Jet - fighter	0.40	0.07

The tail size can be estimated from the tail volume coefficient if the tail lever arms l_H and l_V are known. The lever arms are not, however, fixed until the position of the wing has been established. However, this only takes place. For this reason the tail lever arms can only be estimated from the length of the fuselage in this case.

Table: Conventional tail lever arms of horizontal and vertical tails.

Aircraft configuration	Average of l_h and l_v
Propeller in front of fuselage	60% of fuselage length
Engines on the wing	50 ... 55% of fuselage length
Engines on the tail	45... 50% of fuselage length
Control canard	30 ... 50% of fuselage length
Sailplane	65% of fuselage length

- The tail volume coefficients can be reduced by 10% to 15% in the case of trimmable horizontal stabilizers.
- In the case of a T-tail, the tail volume coefficients can be reduced by 5% for horizontal and vertical tailplane due to the end plate effect and the improved flow.
- In the case of a control canard a tail volume coefficient of 0.1 can be set. In the case of a lifting canard the tail volume coefficient method cannot be applied. Instead a ratio of the areas of canard and wing is established.

- If the criteria for stability and control determine the dimensioning of an aircraft's tail design, the tail volume coefficients can be reduced by approximately 10% if the aircraft has an electronic flight control system, EFCS. However, for transport aircraft other criteria (such as engine failure for the rudder) often determine the dimensioning, so that tail area cannot necessarily be saved through an EFCS.

Functions of Empennage

Empennages create a force that acts upon a lever arm. Consequently a moment is created through empennages:

- The horizontal tailplane creates a moment around the lateral axis (pitch).
- The vertical tailplane (fin) principally creates a moment around the vertical axis (yaw).

Ailerons and spoilers on the wing principally create a moment around the longitudinal axis (roll).

Control surfaces on empennages and on the wing are the customary way to create moments. However, there are other possibilities for creating moments:

- Moving the center of gravity (tail aft aircraft).
- Engine thust (control jets on the VTOL aircraft).

Empennages ensure trim, stability and control. These three aspects are detailed in the next paragraphs.

Trim

The moment created by an empennage balances out moments occurring on the aircraft for another reason. The horizontal tailplane, for example, balances out the wing moment. In the case of propeller aircraft, the rotating slipstream causes a moment at the rear of the fuselage and at the vertical tailplane. The vertical tailplane has to compensate for this moment. If an engine fails on a multi-engine aircraft, the vertical tailplane compensates for an asymmetrical moment distribution around the vertical axis.

The trim has to be guaranteed for all prescribed center-of-gravity positions, airspeeds, flap and landing gear positions and in the event of engine failure.

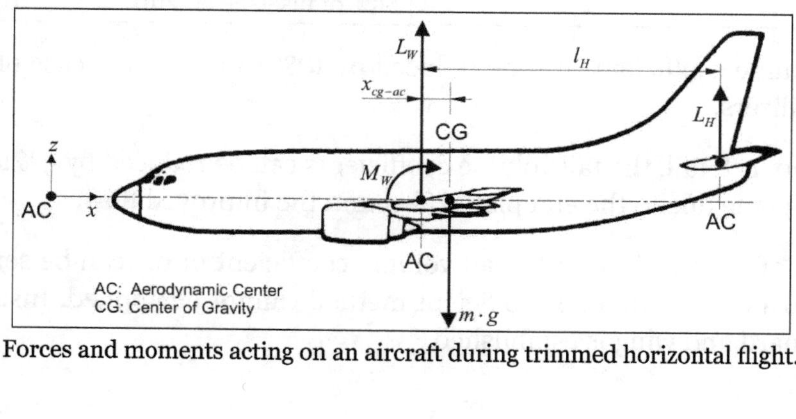

Forces and moments acting on an aircraft during trimmed horizontal flight.

Stability

Stability refers to the capacity of the aircraft to return to the original flying position after a disturbance from outside or after a brief control input. A distinction is made between static stability and dynamic stability.

- Static stability: Longitudinal static stability ensures that the airspeed remains stable.
- Dynamic stability is contingent upon static stability. But an aircraft is not necessarily dynamically stable when it is statically stable, because if the aircraft returns to its original position after a disturbance, it can, of course, easily overshoot the original position. If this oscillation ceases after a while (or an overshoot does not occur), this oscillation of the aircraft is dynamically stable. But if the amplitude of oscillation becomes greater and greater, this oscillation of the aircraft is dynamically instable. Conventional aircraft exhibit the following "oscillation forms" or, to be more precise, modes (it does not always have to be an oscillation; it might also be a heavily damped movement):
    - In a longitudinal movement (i.e. around the lateral axis). Short period mode, phugoid.
    - In a lateral movement (i.e. around the longitudinal and vertical axis). Spiral mode, Dutch roll mode.

The modes can best be explained with a small model aircraft in the hand or in flight. Therefore, a further description is dispensed with at this point.

Control

Critical flight states for the empennage dimensioning from the point of view of control are:

- Horizontal tailplane: critical combination of center-of-gravity position, flap position and airspeed; rotation during take-off; flare when landing: control with trimmed horizontal stabilizer.
- Vertical tailplane (fin): Engine failure in cruise and during take-off and landing. Engine failure during take-off run, landing with crosswind (sideslip to compensate for crosswind component), spinning.

An aircraft must possess sufficient maneuverability in accordance with its flight mission. It is scarcely possible to derive maneuverability criteria from the civil certification regulations. Instead the findings contained in military regulations – also for transport aircraft – are used in the design. In the development phase a simulator model is created and the future aircraft is "flown" and assessed by test pilots. The lever arm and aileron must be large enough for sufficient maneuverability. In addition, it must be possible to deflect the control surfaces quickly enough.

Shapes of the Empennage

Different empennage shapes are shown on selected aircraft in figure. The conventional tail provides appropriate stability and control and also leads to the most lightweight construction in most cases. Approximately 70 % of aircraft are fitted with a conventional tail. Spin characteristics can be bad in the case of a conventional tail due to the blanketing of the vertical tailplane. The downwash of the wing is relatively large in the area of the horizontal tailplane. Rear engines cannot be teamed with conventional tails. Stabilizer trim is possible with comparatively low complexity. A larger vertical tailplane height is more appropriate for a conventional tail than a T-tail.

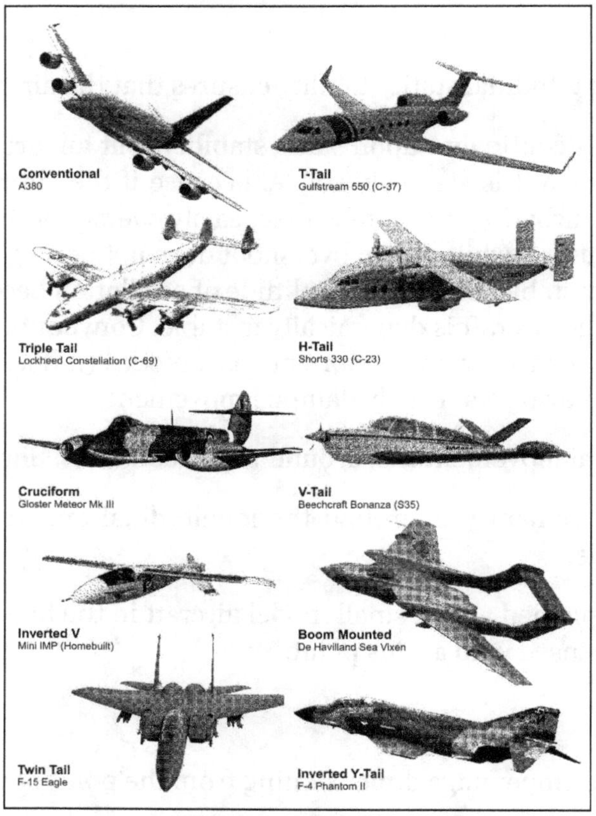

Empennages of conventional aircraft configurations.

The T-tail is heavier than the conventional tail because the vertical tailplane has to support the horizontal tailplane. However, the T-tail has advantages that partly compensate for the described main disadvantage (weight). Owing to the end plate effect, the vertical tailplane can be smaller. The horizontal tailplane is more effective because it is positioned out of the airflow behind the wing and is subjected to less downwash. It can therefore be smaller. For the same reason the horizontal tailplane is also subject to less tail buffeting. The T-tail creates space for engines that are to be placed at the rear. The T-tail looks good, according to general opinion.

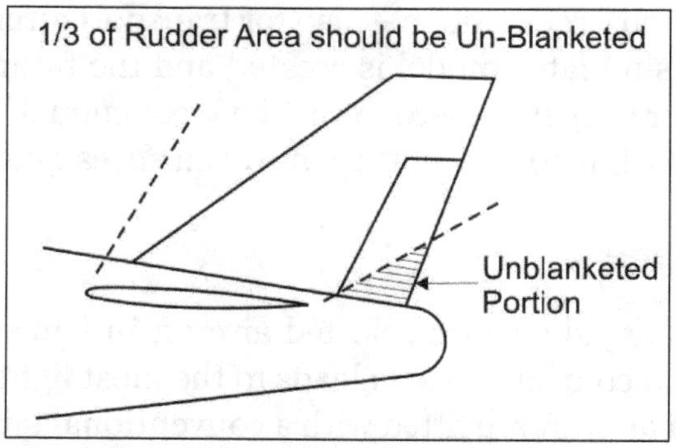

Influence of the empennage design on the spin recovery characteristics.

With T-tails the problem of deep stall must be taken into account. In the case of high angles of attack the horizontal tailplane can be caught up in the airflow behind the wing and be blanketed. If, in addition, the wing tends to make the aircraft pitch up at high angles of attack, a situation may arise in which the aircraft can no longer be recovered from the stall. Figure shows admissible positions of the horizontal tailplane.

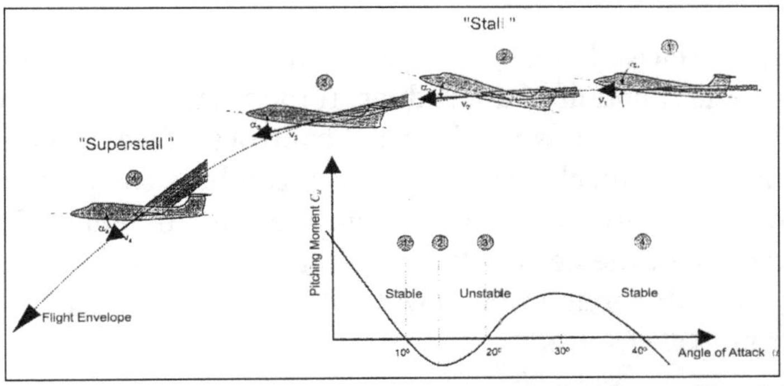

Flight envelope, angle of attack and pitching moment during deep stall and super stall.

The cruciform tail is a compromise between a conventional tail and a T-tail. The cruciform tail weighs less than the T-tail and allows the engines to be placed at the rear (e.g. Caravelle). However, the cruciform tail does not have a surface area advantage due to the end plate effect like the T-tail.

The aim of the V-tail is to achieve a smaller tail area than with horizontal and vertical tailplanes, for example in the form of the conventional tail. The V-tail is designed as follows: In the first step the required areas of a conventional horizontal tailplane S_H and vertical tailplane S_V are determined. Theoretically the V-tail provides efficiency as a horizontal and vertical tailplane, corresponding to the projection of the V-tail in the horizontal and vertical. This theoretical approach gives the necessary V angle for the V-tail:

$$v = \arctan \frac{S_V}{S_H}$$

and the necessary area:

$$S_{V-Tail,theory} = \sqrt{S_H^2 + S_V^2}$$

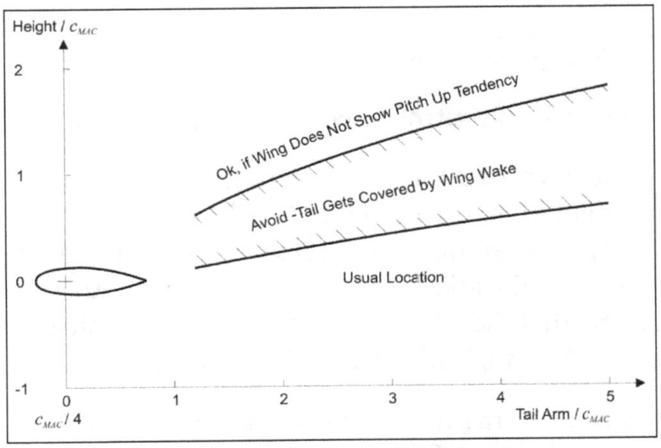

Positioning of horizontal tailplanes.

On the basis of this theoretical analysis the V-tail only requires a tail area of $S_{V-tail} / (S_V + S_H) = 70.7\%$ compared to the conventional tail with $S_V / S_H = 1$. With other S_V / S_H ratios the area saving is less. the V-tail must, however, be larger in practice than the theory suggests for the same efficiency, so that the advantage of the smaller area is lost and a tail area is necessary.

$$S_{V-tail} = S_V + S_H$$

With a V-tail the control surfaces deflect in the same direction in the function of the elevator and in opposite directions in the function of the rudder. If the right rudder pedal is pressed, the right control surface of the V-tail moves down and the left control surface up. One of the disadvantages of the V-tail is the complicated mechanics required to combine the elevator and rudder inputs. Inconveniently a "rudder deflection" of the V-tail causes a roll moment against the desired turn. A roll moment in the direction of the desired turn is, on the other hand, achieved with the inverted V-tail. However, many aircraft configurations will not be able to accommodate an inverted V-tail due to the necessary ground clearance.

A twin tail can be used if a single vertical tailplane would be too big. Twin tails are covered less by the front fuselage in the case of high angles of attack than a vertical tail in the plane of symmetry. For the latter reason twin tails are seen on fighter aircraft that operate in the high angel of attack range. Figure shows additional tail configurations that might be advantageous under certain circumstances.

Other tail features:

- Through the dorsal fin the efficiency of the vertical tailplane where high angles of yaw exist is improved through vortex formation. The stall is thereby moved to higher angles of yaw.
- The ventral fin is not blanketed even with high angles of attack. The ventral fin also serves to prevent lateral instabilities in high-speed flight.

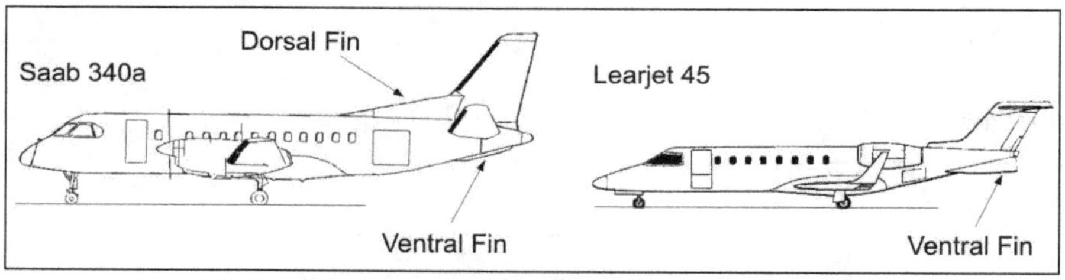

Examples of aircrafts with dorsal fin and ventral fin.

The canard tails are subdivided into control canard and lifting canard:

- In the case of a control canard the wing bears the aircraft's weight. Wings and fuselage alone show neutral stability; the canard is only used for control, but makes the system comprising fuselage, wing and tail instable. An electronic flight control system, EFCS, carries out the regulation and stabilization of the instable aircraft. An aircraft with canard must be designed in such a way that the wing can never be stalled. Instead the canard is first stalled. This necessitates that the wing's lift potential cannot be fully utilized.
- The lifting canard has less drag theoretically because the canard – in contrast to the horizontal tailplane of the tail aft configuration – creates lift (instead of negative lift). By using

the lifting canard the wing must be placed further to the rear. Through this placement the lifting canard is able to facilitate a center-of-gravity range that is normally required. However, the lifting canard displays various disadvantages that restrict its overall utility considerably: the placement of the wing further back on the fuselage increases the nose-heavy moment when using the landing flaps due to the larger lever arm. The wing of the canard must therefore have a greater area with less effective flaps than is customary with the tail aft configuration. Another way of solving this problem is to fit the canard with effective flaps or provide a variable sweep of the canard.

Three-Surface

Wingless

Control-Canard

Tandem Wing

Lifting-Canard

Tailless

- The tandem wing is a lifting canard where the lift forces are approximately evenly distributed between the wing and the canard.
- The three-surface configuration makes it possible to create a pitching moment without influencing the lift on the wing. Therefore it is possible to better optimize the distribution of lift on the wing and thereby reduce the drag. One of the disadvantages is the additional complexity due to an additional area.

All configurations with canards have the disadvantage that the wing lies in a flow disturbed by the empennage placed at the front.

Elevator and Rudder

Elevator and rudder start on the fuselage and extend to approximately 90% of the (semi-) span of the tail, or up to the tip of the tail. They have a chord which accounts for approximately 25 % to 40 % of the chord of the tail. Elevators are deflected downwards by a maximum of 15° to 25° and upwards by a maximum of 25° to 35°. Rudders are deflected by a maximum of 25° to 35°.

Particularly in the case of aircraft with a reversible flight control system, it is important to know the hinge moment required to deflect the rudder in the various flight states. The reason is that the hinge moment determines the hand and foot forces on the flight controls, which may not exceed specific maximum values according to CS 25.143. The hinge moment is calculated with:

$$M_c = \frac{1}{2}\rho V^2 . C_h \cdot S_F \cdot c_F$$

V is the airspeed, S_F is the control surface area, c_F is the rudder depth (measured from the hinge line to the trailing edge). The hinge moment coefficient C_h of a control surface is calculated from the hinge moment derivatives $C_{h\alpha}$ and $C_{h\delta}$. It is important to bear in mind that asymmetrical airfoils already have a hinge moment coefficient C_{ho} at $\alpha = \delta$ o.

$$C_h = C_{h_o} + C_{h_\alpha} \cdot \alpha + C_{h_\delta} \cdot \delta$$

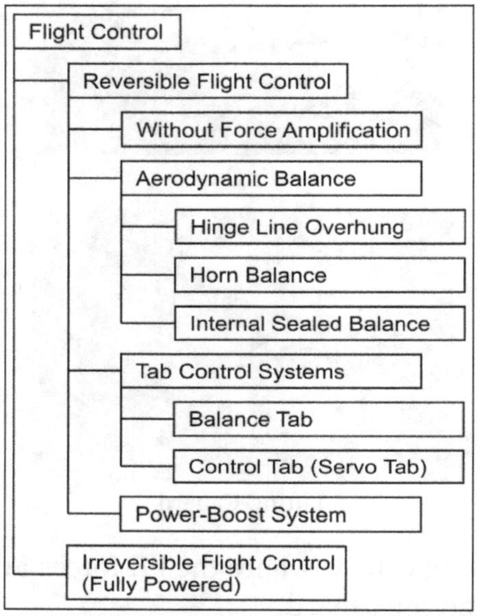

Classification of flight controls and hinge moment reduction possibilities.

According to equation the aerodynamic hinge moment increases with the size and speed of an aircraft. As the control forces may become too large even in small aircraft, measures must be taken to reduce them. The hinge moment is fully or partially carried by the pilot's muscular force on reversible flight controls. On irreversible flight controls the hinge moment is countered by the aircraft's onboard energy systems. Figure shows the main options for reducing control forces. The options are arranged according to increasing effectiveness but also complexity. Figure shows two of these methods for hinge moment reduction. Horn and overhang balance are often applied on small aircraft owing to their simple design.

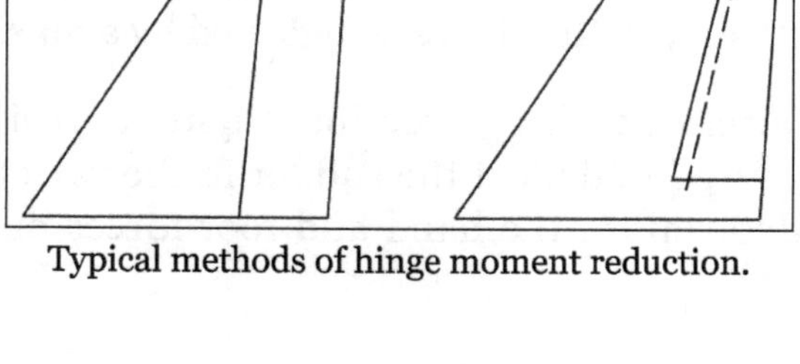

Typical methods of hinge moment reduction.

Flight Control Surfaces

Main Control Surfaces

The main control surfaces of a fixed-wing aircraft are attached to the airframe on hinges or tracks so they may move and thus deflect the air stream passing over them. This redirection of the air stream generates an unbalanced force to rotate the plane about the associated axis.

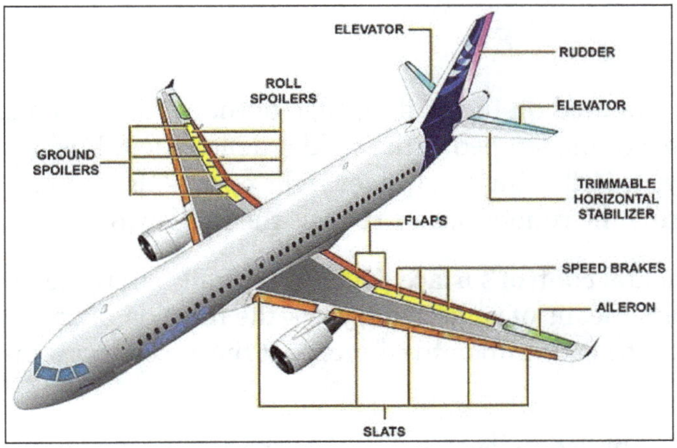

Ailerons

Ailerons are mounted on the trailing edge of each wing near the wingtips and move in opposite directions. When the pilot moves the stick left, or turns the wheel counter-clockwise, the left aileron goes up and the right aileron goes down. A raised aileron reduces lift on that wing and a lowered one increases lift, so moving the stick left causes the left wing to drop and the right wing to rise. This causes the aircraft to roll to the left and begin to turn to the left. Centering the stick returns the ailerons to neutral maintaining the bank angle. The aircraft will continue to turn until opposite aileron motion returns the bank angle to zero to fly straight. The ailerons primarily control roll. Whenever lift is increased, induced drag is also increased. When the stick is moved left to roll the aircraft to the left, the right aileron is lowered which increases lift on the right wing and therefore increases induced drag on the right wing. Using ailerons causes adverse yaw, meaning the nose of the aircraft yaws in a direction opposite to the aileron application. When moving the stick to the left to bank the wings, adverse yaw moves the nose of the aircraft to the *right*. Adverse yaw is more pronounced for light aircraft with long wings, such as gliders. It is counteracted by the pilot with the rudder. Differential ailerons are ailerons which have been rigged such that the down going aileron deflects less than the upward-moving one, reducing adverse yaw.

Elevator

The elevator is a moveable part of the horizontal stabilizer, hinged to the back of the fixed part of the horizontal tail. The elevators move up and down together. When the pilot pulls the stick backward, the elevators go up. Pushing the stick forward causes the elevators to go down. Raised elevators push down on the tail and cause the nose to pitch up. This makes the wings fly at a higher angle of attack, which generates more lift and more drag. Centering the stick returns the elevators to neutral and stops the change of pitch. Some aircraft, such as an MD-80, use a servo tab within the elevator surface to aerodynamically move the main surface into position. The direction of travel of the control tab will thus be in a direction opposite to the main control surface. It is for this reason that an MD-80 tail looks like it has a 'split' elevator system.

In the canard arrangement, the elevators are hinged to the rear of a foreplane and move in the opposite sense, for example when the pilot pulls the stick back the elevators go down to increase the lift at the front and lift the nose up.

Rudder

The rudder is typically mounted on the trailing edge of the vertical stabilizer, part of the empennage. When the pilot pushes the left pedal, the rudder deflects left. Pushing the right pedal causes the rudder to deflect right. Deflecting the rudder right pushes the tail left and causes the nose to yaw to the right. Centering the rudder pedals returns the rudder to neutral and stops the yaw.

The rudder is a fundamental control surface which is typically controlled by pedals rather than at the stick. It is the primary means of controlling yaw—the rotation of an airplane about its vertical axis. The rudder may also be called upon to counter-act the adverse yaw produced by the roll-control surfaces.

If rudder is continuously applied in level flight the aircraft will yaw initially in the direction of the applied rudder – the primary effect of rudder. After a few seconds the aircraft will tend to bank in the direction of yaw.

This arises initially from the increased speed of the wing opposite to the direction of yaw and the reduced speed of the other wing. The faster wing generates more lift and so rises, while the other wing tends to go down because of generating less lift. Continued application of rudder sustains rolling tendency because the aircraft flying at an angle to the airflow – skidding towards the forward wing. When applying right rudder in an aircraft with dihedral the left hand wing will have increased angle of attack and the right hand wing will have decreased angle of attack which will result in a roll to the right. An aircraft with anhedral will show the opposite effect. This effect of the rudder is commonly used in model aircraft where if sufficient diheral or polyhedral is included in the wing design, primary roll control such as ailerons may be omitted altogether.

Secondary Control Surfaces

Spoilers

On low drag aircraft such as sailplanes, spoilers are used to disrupt airflow over the wing and greatly reduce lift. This allows a glider pilot to lose altitude without gaining excessive airspeed. Spoilers are sometimes called "lift dumpers". Spoilers that can be used asymmetrically are called spoilerons and can affect an aircraft's roll.

Flaps

Flaps are mounted on the trailing edge on the inboard section of each wing (near the wing roots). They are deflected down to increase the effective curvature of the wing. Flaps raise the maximum lift coefficient of the aircraft and therefore reduce its stalling speed. They are used during low speed, high angle of attack flight including take-off and descent for landing. Some aircraft are equipped with "flaperons", which are more commonly called "inboard ailerons". These devices function primarily as ailerons, but on some aircraft, will "droop" when the flaps are deployed, thus acting as both a flap and a roll-control inboard aileron.

Slats

Slats, also known as leading edge devices, are extensions to the front of a wing for lift augmentation, and are intended to reduce the stalling speed by altering the airflow over the wing. Slats may be fixed or retractable – fixed slats give excellent slow speed and STOL capabilities, but compromise higher speed performance. Retractable slats, as seen on most airliners, provide reduced stalling speed for take-off and landing, but are retracted for cruising.

Air Brakes

Air brakes are used to increase drag. Spoilers might act as air brakes, but are not pure air brakes as they also function as lift-dumpers or in some cases as roll control surfaces. Air brakes are usually surfaces that deflect outwards from the fuselage (in most cases symmetrically on opposing sides) into the airstream in order to increase form-drag. As they are in most cases located elsewhere on the aircraft, they do not directly affect the lift generated by the wing. Their purpose is to slow down the aircraft. They are particularly useful when a high rate of descent is required. They are common on high performance military aircraft as well as civilian aircraft, especially those lacking reverse thrust capability.

Tabs

The force of the air against a control surface during the high speed of flight can make it difficult to move and hold that control surface in the deflected position. A control surface might also be too sensitive for similar reasons. Several different tabs are used to aid with these types of problems. The table in figure below summarizes the various tabs and their uses.

Flight Control Tabs			
Type	**Direction of Motion (in relation to control surface)**	**Activation**	**Effect**
Trim	Opposite	Set by pilot from cockpit. Uses independent linkage.	Statically balances the aircraft in flight. Allows "hands off" maintenance of flight condition.
Balance	Opposite	Moves when pilot moves control surface. Coupled to control surface linkage.	Aids pilot in overcoming the force needed to move the control surface.
Servo	Opposite	Directly linked to flight control input device. Can be primary or back-up means of control.	Aerodynamically positions control surfaces that require too much force to move manually.
Anti-balance or Anti-servo	Same	Directly linked to flight control input device.	Increases force needed by pilot to change flight control position. De-sensitizes flight controls.
Spring	Opposite	Located in line of direct linkage to servo tab. Spring assists when control forces become too high in high-speed flight.	Enables moving control surface when forces are high. Inactive during slow flight.

While in flight, it is desirable for the pilot to be able to take his or her hands and feet off of the controls and have the aircraft maintain its flight condition. Trims tabs are designed to allow this. Most trim tabs are small movable surfaces located on the trailing edge of a primary flight control surface. A small movement of the tab in the direction opposite of the direction the flight control surface is deflected, causing air to strike the tab, in turn producing a force that aids in maintaining the flight control surface in the desired position. Through linkage set from the cockpit, the tab can be positioned so that it is actually holding the control surface in position rather than the pilot. Therefore, elevator tabs are used to maintain the speed of the aircraft since they assist in maintaining the selected pitch. Rudder tabs can be set to hold yaw in check and maintain heading. Aileron tabs can help keep the wings level.

Occasionally, a simple light aircraft may have a stationary metal plate attached to the trailing edge of a primary flight control, usually the rudder. This is also a trim tab as shown in figure. It can be bent slightly on the ground to trim the aircraft in flight to a hands-off condition when flying straight and level. The correct amount of bend can be determined only by flying the aircraft after an adjustment. Note that a small amount of bending is usually sufficient.

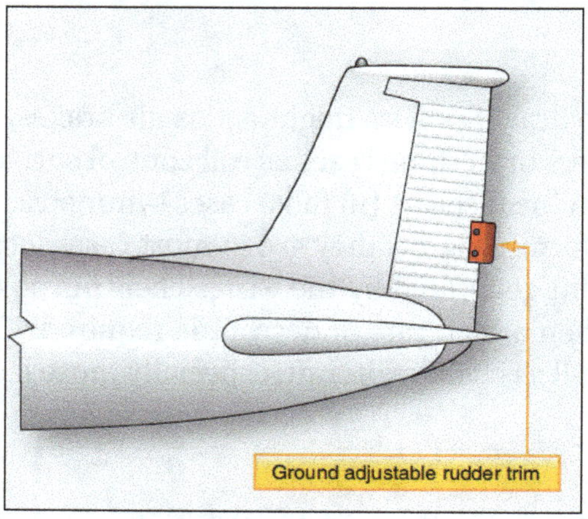

The aerodynamic phenomenon of moving a trim tab in one direction to cause the control surface to experience a force moving in the opposite direction is exactly what occurs with the use of balance tabs. Often, it is difficult to move a primary control surface due to its surface area and the speed of the air rushing over it. Deflecting a balance tab hinged at the trailing edge of the control surface in the opposite direction of the desired control surface movement causes a force to position the surface in the proper direction with reduced force to do so. Balance tabs are usually linked directly to the control surface linkage so that they move automatically when there is an input for control surface movement. They also can double as trim tabs, if adjustable in the flight deck.

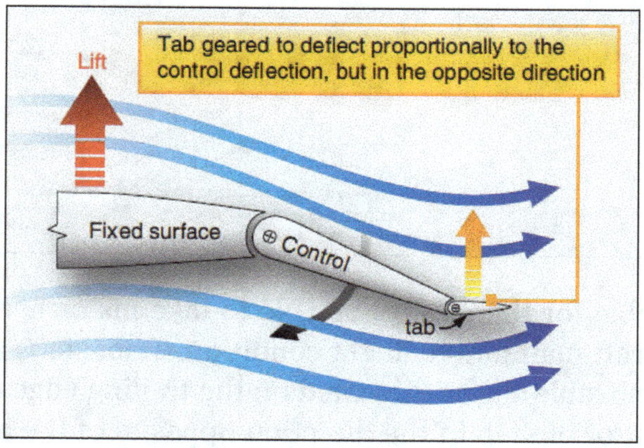

A servo tab is similar to a balance tab in location and effect, but it is designed to operate the primary flight control surface, not just reduce the force needed to do so. It is usually used as a means to back up the primary control of the flight control surfaces.

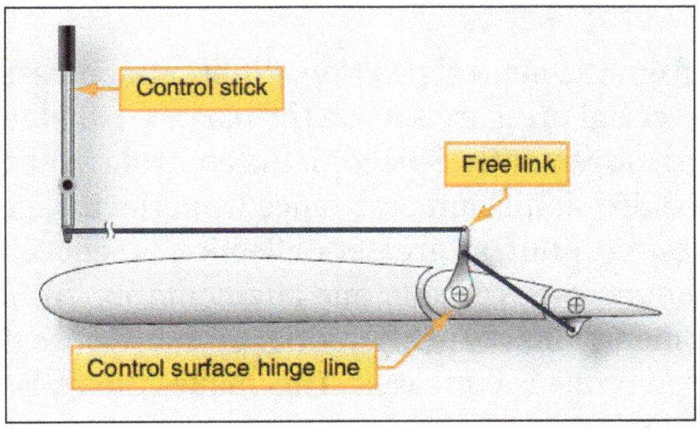

On heavy aircraft, large control surfaces require too much force to be moved manually and are usually deflected out of the neutral position by hydraulic actuators. These power control units are signaled via a system of hydraulic valves connected to the yoke and rudder pedals. On fly-by-wire aircraft, the hydraulic actuators that move the flight control surfaces are signaled by electric input. In the case of hydraulic system failure(s), manual linkage to a servo tab can be used to deflect it. This, in turn, provides an aerodynamic force that moves the primary control surface. A control surface may require excessive force to move only in the final stages of travel. When this is the case, a spring tab can be used. This is essentially a servo tab that does not activate until an effort is made to move the control surface beyond a certain point. When reached, a spring in line of the control linkage aids in moving the control surface through the remainder of its travel.

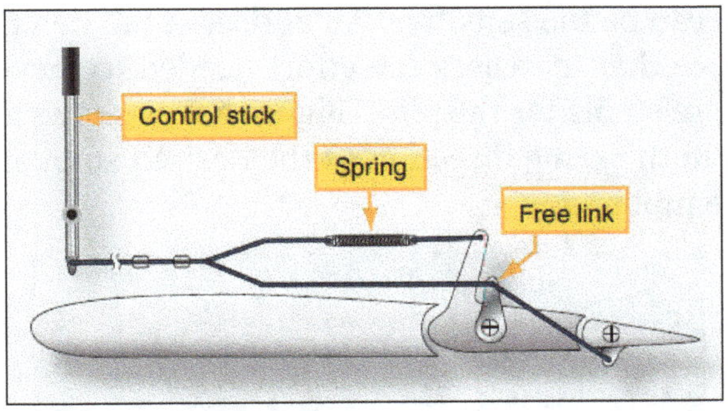

Figure shows another way of assisting the movement of an aileron on a large aircraft. It is called an aileron balance panel. Not visible when approaching the aircraft, it is positioned in the linkage that hinges the aileron to the wing.

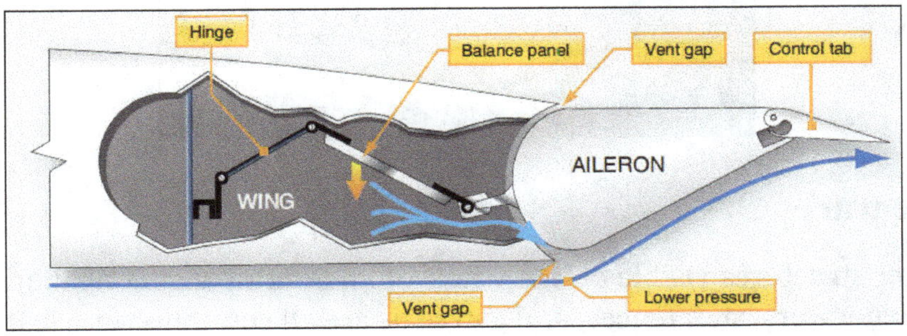

Balance panels have been constructed typically of aluminum skin-covered frame assemblies or aluminum honeycomb structures. The trailing edge of the wing just forward of the leading edge of the aileron is sealed to allow controlled airflow in and out of the hinge area where the balance panel is located.

When the aileron is moved from the neutral position, differential pressure builds up on one side of the balance panel. This differential pressure acts on the balance panel in a direction that assists the aileron movement. For slight movements, deflecting the control tab at the trailing edge of the aileron is easy enough to not require significant assistance from the balance tab. (Moving the control tab moves the ailerons as desired.) But, as greater deflection is requested, the force resisting control tab and aileron movement becomes greater and augmentation from the balance tab is needed. The seals and mounting geometry allow the differential pressure of airflow on the balance panel to increase as deflection of the ailerons is increased. This makes the resistance felt when moving the aileron controls relatively constant.

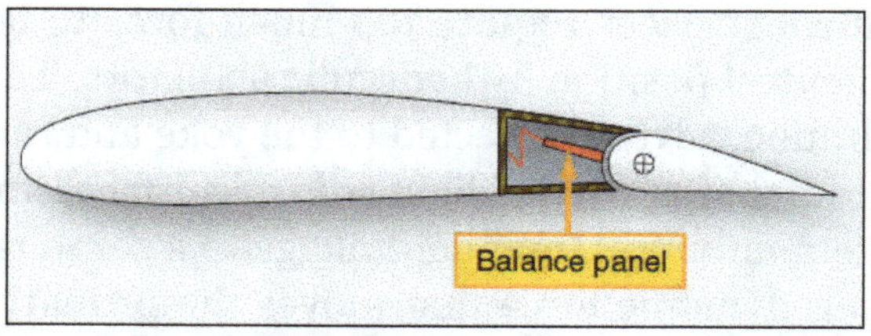

The trailing edge of the wing just forward of the leading edge of the aileron is sealed to allow controlled airflow in and out of the hinge area where the balance panel is located.

Antiservo tabs, as the name suggests, are like servo tabs but move in the same direction as the primary control surface. On some aircraft, especially those with a movable horizontal stabilizer, the input to the control surface can be too sensitive. An antiservo tab tied through the control linkage creates an aerodynamic force that increases the effort needed to move the control surface. This makes flying the aircraft more stable for the pilot. Figure below shows an antiservo tab in the near neutral position. Deflected in the same direction as the desired stabilator movement, it increases the required control surface input.

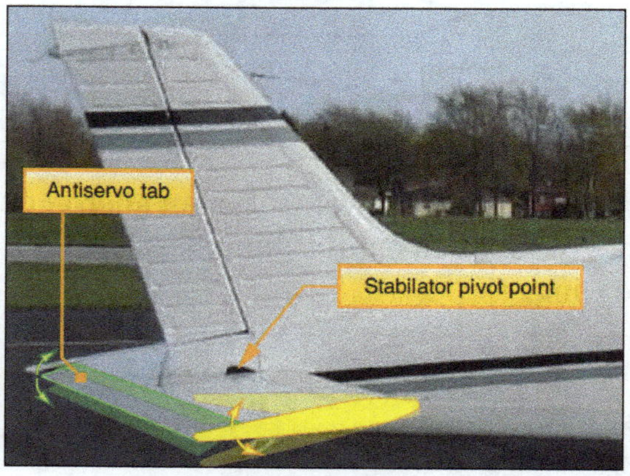

Other Wing Features

There may be other structures visible on the wings of an aircraft that contribute to performance. Winglets, vortex generators, stall fences, and gap seals are all common wing features.

A winglet is an obvious vertical upturn of the wing's tip resembling a vertical stabilizer. It is an aerodynamic device designed to reduce the drag created by wing tip vortices in flight. Usually made from aluminum or composite materials, winglets can be designed to optimize performance at a desired speed.

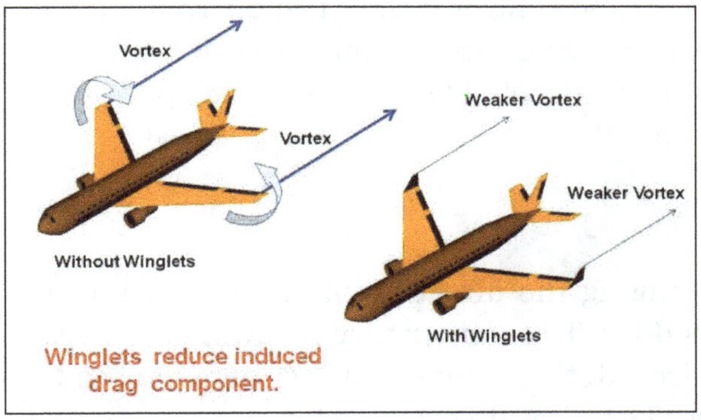

Vortex generators are small airfoil sections usually attached to the upper surface of a wing. They are designed to promote positive laminar airflow over the wing and control surfaces. Usually made of aluminum and installed in a spanwise line or lines, the vortices created by these devices swirl downward assisting maintenance of the boundary layer of air flowing over the wing.

A chordwise barrier on the upper surface of the wing, called a stall fence, is used to halt the spanwise flow of air. During low speed flight, this can maintain proper chordwise airflow reducing the tendency for the wing to stall. Usually made of aluminum, the fence is a fixed structure most common on swept wings, which have a natural spanwise tending boundary air flow.

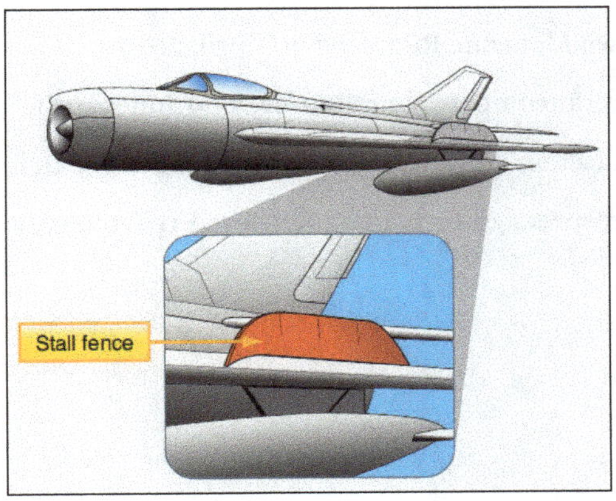

Often, a gap can exist between the stationary trailing edge of a wing or stabilizer and the movable control surface(s). At high angles of attack, high pressure air from the lower wing surface can be disrupted at this gap. The result can be turbulent airflow, which increases drag. There is also a tendency for some lower wing boundary air to enter the gap and disrupt the upper wing surface airflow, which in turn reduces lift and control surface responsiveness. The use of gap seals is common to promote smooth airflow in these gap areas. Gap seals can be made of a wide variety of materials ranging from aluminum and impregnated fabric to foam and plastic. Figure below shows some gap seals installed on various aircraft.

Turning the Aircraft

Unlike turning a boat, changing the direction of an aircraft normally must be done with the ailerons rather than the rudder. The rudder turns (yaws) the aircraft but has little effect on its direction of travel. With aircraft, the change in direction is caused by the horizontal component of lift, acting on the wings. The pilot tilts the lift force, which is perpendicular to the wings, in the direction of the intended turn by rolling the aircraft into the turn. As the bank angle is increased, the lifting force can be split into two components: one acting vertically and one acting horizontally.

If the total lift is kept constant, the vertical component of lift will decrease. As the weight of the aircraft is unchanged, this would result in the aircraft descending if not countered. To maintain level flight requires increased positive (up) elevator to increase the angle of attack, increase the total lift generated and keep the vertical component of lift equal with the weight of the aircraft. This cannot continue indefinitely. The total load factor required to maintain level flight is directly related to the bank angle. This means that for a given airspeed, level flight can only be maintained up to a certain given angle of bank. Beyond this angle of bank, the aircraft will suffer an accelerated stall if the pilot attempts to generate enough lift to maintain level flight.

References

- The-three-stages-of-aircraft-design: monroeaerospace.com, Retrieved 8 January, 2019
- Aircraft-design-characteristics-part-one: flightliteracy.com, Retrieved 13 May, 2019
- Aircraft-design-characteristics-part-two: flightliteracy.com, Retrieved 25 February, 2019
- New-materials-being-used-in-aircraft-industry: wordpress.com, Retrieved 16 January, 2019
- Composite-materials-aircraft-structure-282777: thebalancecareers.com, Retrieved 29 March, 2019
- Aviation:cockpit-airplane: wikidot.com, Retrieved 30 April, 2019
- Types-of-aircraft-wings: aircraftcompare.com, Retrieved 29 June, 2019
- Empennage: flight-mechanic.com, Retrieved 14 July, 2019
- Flight-control-surfaces: aerospacengineering.net, Retrieved 11 January, 2019

3

Use of Aerodynamics in Aircrafts

The branch of science which focuses on the motion of air and its interaction with solid objects is termed as aerodynamics. Some of the areas of study within aerodynamics in relation to aircraft are effect of wing planform, the angle of attack and the thrust on aircraft. These diverse applications of aerodynamics in aircraft have been thoroughly discussed in this chapter.

Aerodynamics

Aerodynamics is the science of moving air and its impact on solid bodies placed in the flow field as an obstacle. Being a sub-field, most of the equations from fluid dynamics apply to aerodynamics as well, including all the governing equations, turbulence, boundary layer theory, and ideal gas assumption.

Flow Classification and Equations

An aerodynamic flow field can be classified based on the following physical parameters: velocity, pressure, and temperature. Additionally, density and viscosity can be used as well for classification. The most common flow classification practices are velocity and pressure-based.

Velocity-based Classification

Subsonic flows are flow fields in which the air speed does not exceed the local speed of sound, that is $M<1$. Typical examples are common ground vehicles like cars, trains, F1 cars, and sport airplanes. MM represents the Mach number which is the flow velocity past a boundary.

Transonic flows include both subsonic and supersonic flow regimes $0.8<M<1.2$. Typically, commercial airplanes fly in this speed region. A supersonic flow zone appears when the otherwise subsonic fluid accelerates above sound speed due to surface curvatures (e.g. suction side of the wing).

Supersonic flows are defined as having velocities greater than the speed of sound everywhere in the domain ($M>1$). Military jets and gun bullets are good examples for this velocity.

We could also include hypersonic flows, where the speed is much higher than the speed of sound. There is no accurate definition of this value ($M>>1$). Some hypersonic rocket-powered airplanes and spacecraft's can be mentioned as examples.

Density/Pressure-based Classification

Incompressible flow refers to a flow where the material (air) density remains constant. Under Mach 0.3 this is a good assumption for industrial flows and helps engineers make the calculation/simulation simpler. Please note that an incompressible flow does not mean that the fluid has to be incompressible itself.

A flow can be considered as compressible if the density changes along a streamline. Generally, this is the case for subsonic flows above Mach 0.3. Transonic, supersonic, and hypersonic flows are all compressible.

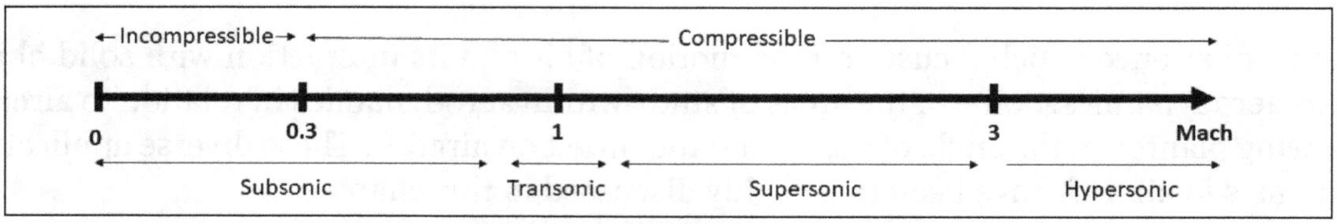

Velocity and Compressibility of flows.

Effect of Wing Planform

Understanding the effects of different wing planforms is important when learning about wing performance and airplane flight characteristics. A planform is the shape of the wing as viewed from directly above and deals with airflow in three dimensions. Aspect ratio, taper ratio, and sweepback are factors in planform design that are very important to the overall aerodynamic characteristic of a wing.

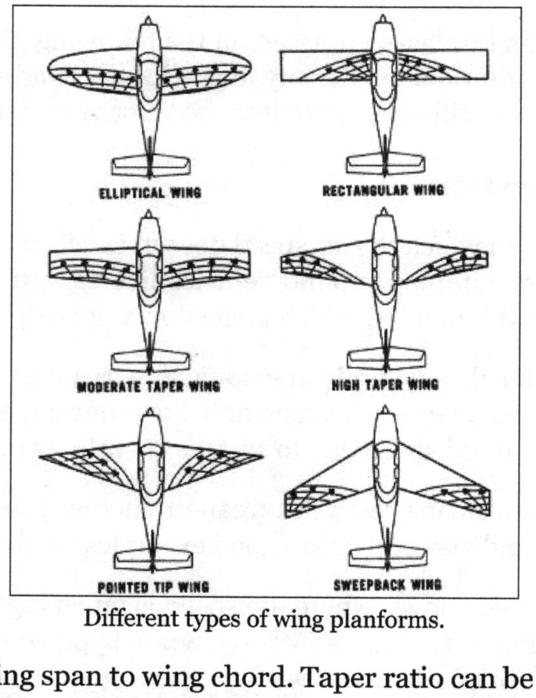

Different types of wing planforms.

Aspect ratio is the ratio of wing span to wing chord. Taper ratio can be either in planform or thickness, or both. In its simplest terms, it is a decrease from wing root to wingtip in wing chord or wing thickness. Sweepback is the rearward slant of a wing, horizontal tail, or other airfoil surface.

There are two general means by which the designer can change the planform of a wing and both will affect the aerodynamic characteristics of the wing. The first is to effect a change in the aspect ratio. Aspect ratio is the primary factor in determining the three dimensional characteristics of the ordinary wing and its lift/drag ratio. An increase in aspect ratio with constant velocity will decrease the drag, especially at high angles of attack, improving the performance of the wing when in a climbing attitude.

A decrease in aspect ratio will give a corresponding increase in drag. It should be noted, however, that with an increase in aspect ratio there is an increase in the length of span, with a corresponding increase in the weight of the wing structure, which means the wing must be heavier to carry the same load. For this reason, part of the gain (due to a decrease in drag) is lost because of the increased weight, and a compromise in design is necessary to obtain the best results from these two conflicting conditions.

The second means of changing the planform is by tapering (decreasing the length of chord from the root to the tip of the wing). In general, tapering causes a decrease in drag (most effective at high speeds) and an increase in lift. There is also a structural benefit due to a saving in weight of the wing.

Most training and general aviation type airplanes are operated at high coefficients of lift, and therefore require comparatively high aspect ratios. Airplanes that are developed to operate at very high speeds demand greater aerodynamic cleanness and greater strength, which require low aspect ratios. Very low aspect ratios result in high wing loadings and high stall speeds. When sweepback is combined with low aspect ratio, it results in flying qualities very different from a more conventional high aspect ratio airplane configuration. Such airplanes require very precise and professional flying techniques, especially at slow speeds, while airplanes with a high aspect ratio are usually more forgiving of improper pilot techniques.

The elliptical wing is the ideal subsonic planform since it provides for a minimum of induced drag for a given aspect ratio, though as we shall see, its stall characteristics in some respects are inferior to the rectangular wing. It is also comparatively difficult to construct. The tapered airfoil is desirable from the standpoint of weight and stiffness, but again is not as efficient aerodynamically as the elliptical wing. In order to preserve the aerodynamic efficiency of the elliptical wing, rectangular and tapered wings are sometimes tailored through use of wing twist and variation in airfoil sections until they provide as nearly as possible the elliptical wing's lift distribution. While it is true that the elliptical wing provides the best coefficients of lift before reaching an incipient stall, it gives little advance warning of a complete stall, and lateral control may be difficult because of poor aileron effectiveness.

In comparison, the rectangular wing has a tendency to stall first at the wing root and provides adequate stall warning, adequate aileron effectiveness, and is usually quite stable. It is, therefore, favored in the design of low cost, low speed airplanes.

Angle of Attack

Angle of Attack (AOA) is used to define the angle between the wing chord line and the flight path. This is not to be confused with the relation of the aircraft to the Earth's surface. This is called the attitude and is seldom, if ever, the same as the angle of attack.

When the angle is small, the aircraft is said to be at a low angle of attack. When the angle is large, the aircraft is said to be at a high angle of attack. During landing, an airplane may have a level attitude, but a high angle of attack, because the flight path is downward and the approaching wind is parallel to the flight path. During climb, an airplane can be in a nose-high attitude, but at a low angle of attack.

Two variables can change the amount of lift generated by a wing in a given configuration:

- The speed of air flowing over the airfoil.
- The angle of attack.

An increase in speed or the angle of attack will increase both lift and drag. In level flight, lift must equal the weight of the aircraft. If an aircraft weighs 2,000 lbs., the wing must generate 2,000 lbs. of lift. The speed and angle of attack are interchangeable to a point—therefore, for every airspeed; there is a corresponding angle of attack that will produce the same amount of lift.

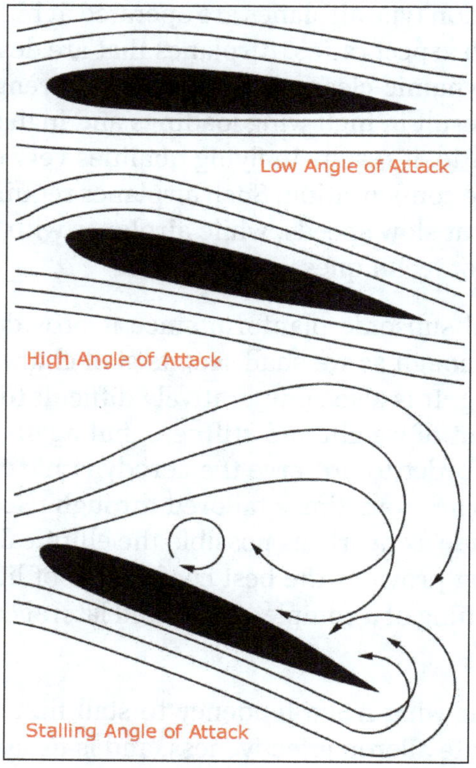

In order for a wing to produce lift, the air flowing past an aircraft, must be aligned to the airfoil in order to provide a smooth airflow. As a wing increases its angle of attack, airflow can no longer flow smoothly over the wing and eddies or *burbles* will form, causing the wing to approach its stall speed. When a wing finally stalls, it will no longer produce lift and with weight unopposed by lift, the aircraft will drop towards the ground. With sufficient altitude, stall recovery can be obtained by decreasing the angle of attack.

An airplane can be stalled at any attitude and at any airspeed such as pulling out of a dive too abruptly or if the airplane is in a steep turn at a high angle of attack, even though the airspeed is high. During dives or turns, centrifugal force will increase *load factors* and if excessive, this will cause the wing to stall.

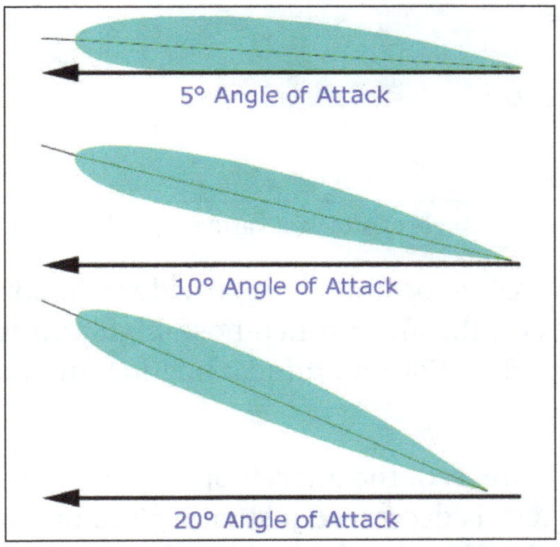

Angle of Attack is the angle between the wing chord line and the flight path.

Thrust on Aircraft

For an aircraft to start moving, thrust must be exerted and be greater than drag. The aircraft continues to move and gain speed until thrust and drag are equal. In order to maintain a constant airspeed, thrust and drag must remain equal, just as lift and weight must be equal to maintain a constant altitude. If in level flight, the engine power is reduced, the thrust is lessened, and the aircraft slows down. As long as the thrust is less than the drag, the aircraft continues to decelerate. To a point, as the aircraft slows down, the drag force will also decrease. The aircraft will continue to slow down until thrust again equals drag at which point the airspeed will stabilize.

Likewise, if the engine power is increased, thrust becomes greater than drag and the airspeed increases. As long as the thrust continues to be greater than the drag, the aircraft continues to accelerate. When drag equals thrust, the aircraft flies at a constant airspeed.

Straight-and-level flight may be sustained at a wide range of speeds. The pilot coordinates AOA and thrust in all speed regimes if the aircraft is to be held in level flight. An important fact related to the principal of lift (for a given airfoil shape) is that lift varies with the AOA and airspeed. Therefore, a large AOA at low airspeeds produces an equal amount of lift at high airspeeds with a low AOA. The speed regimes of flight can be grouped in three categories: low speed flight, cruising flight, and high-speed flight.

When the airspeed is low, the AOA must be relatively high if the balance between lift and weight is to be maintained. If thrust decreases and airspeed decreases, lift will become less than weight and the aircraft will start to descend. To maintain level flight, the pilot can increase the AOA an amount that generates a lift force again equal to the weight of the aircraft. While the aircraft will be flying more slowly, it will still maintain level flight. The AOA is adjusted to maintain lift equal weight. The airspeed will naturally adjust until drag equals thrust and then maintain that airspeed (assumes the pilot is not trying to hold an exact speed).

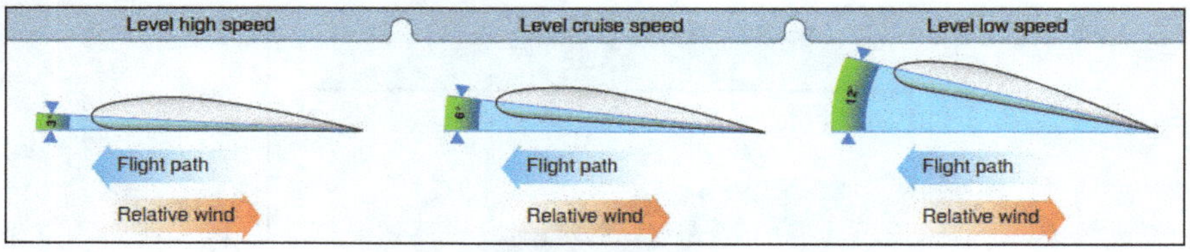

Angle of attack at various speeds.

Straight-and-level flight in the slow-speed regime provides some interesting conditions relative to the equilibrium of forces. With the aircraft in a nose-high attitude, there is a vertical component of thrust that helps support it. For one thing, wing loading tends to be less than would be expected.

In level flight, when thrust is increased, the aircraft speeds up and the lift increases. The aircraft will start to climb unless the AOA is decreased just enough to maintain the relationship between lift and weight. The timing of this decrease in AOA needs to be coordinated with the increase in thrust and airspeed. Otherwise, if the AOA is decreased too fast, the aircraft will descend, and if the AOA is decreased too slowly, the aircraft will climb.

As the airspeed varies due to thrust, the AOA must also vary to maintain level flight. At very high speeds and level flight, it is even possible to have a slightly negative AOA. As thrust is reduced and airspeed decreases, the AOA must increase in order to maintain altitude. If speed decreases enough, the required AOA will increase to the critical AOA. Any further increase in the AOA will result in the wing stalling. Therefore, extra vigilance is required at reduced thrust settings and low speeds so as not to exceed the critical angle of attack. If the airplane is equipped with an AOA indicator, it should be referenced to help monitor the proximity to the critical AOA.

Some aircraft have the ability to change the direction of the thrust rather than changing the AOA. This is accomplished either by pivoting the engines or by vectoring the exhaust gases.

Some aircraft have the ability to change the direction of thrust.

Drag on Aircraft

In aerodynamics, drag is defined as the force that opposes forward motion through the atmosphere and is parallel to the direction of the free-stream velocity of the airflow. Drag must be overcome by thrust in order to achieve forward motion.

Drag is generated by nine conditions associated with the motion of air particles over the aircraft. There are several types of drag: form, pressure, skin friction, parasite, induced, and wave.

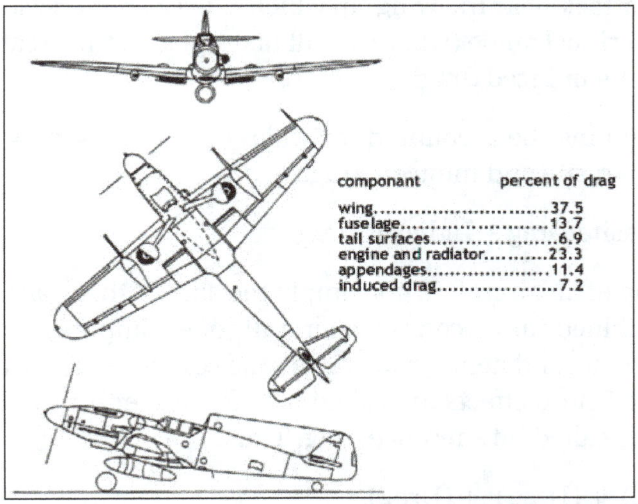

Me-109G German fighter from World War II. Shown is the percentage breakdown of the drag (includes interference drag) of the components.

The term "separation" refers to the smooth flow of air as it closely hugs the surface of the wing then suddenly breaking free of the surface and creating a chaotic flow. The second picture shows examples of air flowing past a variety of objects. The bottom shows well behaved, laminar flow (flow in layers) where the flow stays attached (close to the surface) of the object. The object just above has a laminar flow for the first half of the object and then the flow begins to separate from the surface and form many chaotic tiny vortex flows called vortices. The two objects just above them have a large region of separated flow. The greater the region of separated flow, the greater the drag. This is why airplane designers go to such effort to streamline wings and tails and fuselages — to minimize drag.

Induced Drag

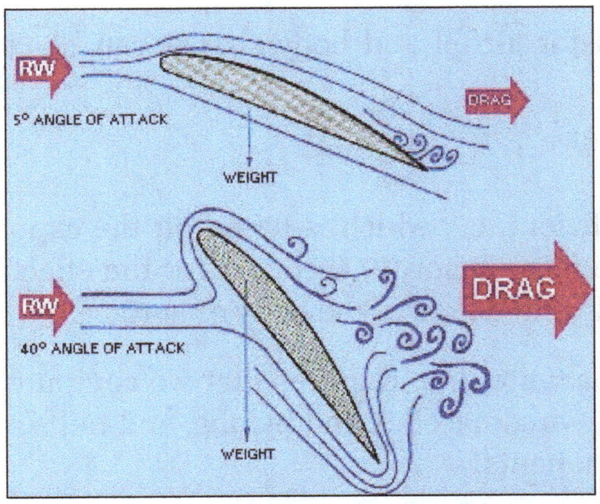

Induced drag is a by-product of lift.

Induced drag is the drag created by the vortices at the tip of an aircraft's wing. Induced drag is the drag due to lift. The high pressure underneath the wing causes the airflow at the tips of the wings to curl around from bottom to top in a circular motion. This results in a trailing vortex. Induced drag increases in direct proportion to increases in the angle of attack. The circular motion creates a change in the angle of attack near the wing tip which causes an increase in drag. The greater the angle of attack up to the critical angle (where a stall takes place), the greater the amount of lift developed and the greater the induced drag.

All of these types of drag must be accounted for when determining drag for subsonic flight. The total drag is the sum of parasite and induced drag.

Total Drag = Parasite Drag + Induced Drag

But the net (or total) drag of an aircraft is not simply the sum of the drag of its components. When the components are combined into a complete aircraft, one component can affect the air flowing around and over the airplane, and hence, the drag of one component can affect the drag associated with another component. These effects are called interference effects, and the change in the sum of the component drags is called interference drag. Thus,

$$(\text{Drag})_{1+2} = (\text{Drag})_1 + (\text{Drag})_2 + (\text{Drag})_{\text{interference}}$$

Generally, interference drag will add to the component drags but in a few cases, for example, adding tip tanks to a wing, total drag will be less than the sum of the two component drags because of the reduction of induced drag.

Parasite Drag

The parasite drag of a typical airplane in the cruise configuration consists primarily of the skin friction, roughness, and pressure drag of the major components. There is usually some additional parasite drag due to such things as fuselage upsweep, control surface gaps, base areas, and other extraneous items. Since most of the elements that make up the total parasite drag are dependent on Reynolds number and since some are dependent on Mach number, it is necessary to specify the conditions under which the parasite drag is to be evaluated. In the method of these notes, the conditions selected are the Mach number and the Reynolds number corresponding to the flight condition of interest.

The basic parasite drag area for airfoil and body shapes can be computed from the following: expression:

$$f = k\, c_f\, S_{wet}$$

where, the skin friction coefficient, c_f , which is based on the exposed wetted area includes the effects of roughness, and the form factor, k, accounts for the effects of both supervelocities and pressure drag. S_{wet} is the total wetted area of the body or surface.

Computation of the overall parasite drag requires that we compute the drag area of each of the major components (fuselage, wing, nacelles and pylons, and tail surfaces) and then evaluate the additional parasite drag components.

We thus write:

$$C_{D_p} = S\, k_i c_{f_i} S_{weti} \,/\, S_{ref} + C_{Dupsweep} + C_{Dgap} + C_{Dnac\_base} + C_{Dmisc}$$

Where, the first term includes skin friction, and pressure drag at zero lift of the major components. c_{f_i} is the average skin friction coefficient for a rough plate with transition at flight Reynolds number. Equivalent roughness is determined from flight test data.

Form Drag

Form drag and pressure drag are virtually the same type of drag. Form or pressure drag is caused by the air that is flowing over the aircraft or airfoil. The separation of air creates turbulence and results in pockets of low and high pressure that leave a wake behind the airplane or airfoil (thus the name pressure drag). This opposes forward motion and is a component of the total drag. Since this drag is due to the shape, or form of the aircraft, it is also called form drag. Streamlining the aircraft will reduce form drag, and parts of an aircraft that do not lend themselves to streamlining are enclosed in covers called fairings, or a cowling for an engine, that have a streamlined shape. Airplane components that produce form drag include (1) the wing and wing flaps, (2) the fuselage, (3) tail surfaces, (4) nacelles, (5) landing gear, (6) wing tanks and external stores, and (7) engines.

Skin friction drag is caused by the actual contact of the air particles against the surface of the aircraft. This is the same as the friction between any two objects or substances. Because skin friction drag is an interaction between a solid (the airplane surface) and a gas (the air), the magnitude of skin friction drag depends on the properties of both the solid and the gas. For the solid airplane, skin fiction drag can be reduced, and airspeed can be increased somewhat, by keeping an aircraft's surface highly polished and clean. For the gas, the magnitude of the drag depends on the viscosity of the air. Along the solid surface of the airplane, a boundary layer of low energy flow is generated. The magnitude of the skin friction depends on the state of this flow.

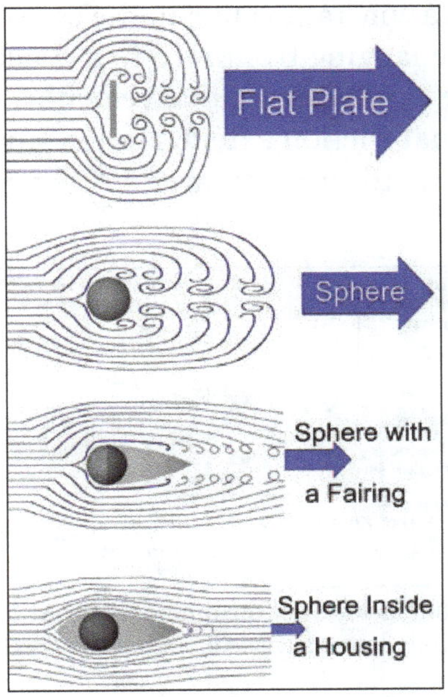

Skin Friction

An important aerodynamic force during low-speed subsonic flight is the shear force (the sideways force or internal friction) caused by viscous airflow over the surfaces of the vehicle. This shear force is referred to as the skin-friction force or skin-friction drag and depends strongly on the Reynolds number, surface roughness, and pressure gradients. In addition to the pressure forces that act everywhere perpendicular to (normal to) a body in moving air, viscous forces are also present. It is these viscous forces that modify the lift that would exist under ideal conditions (air is inviscid and incompressible) and help create the real drag.

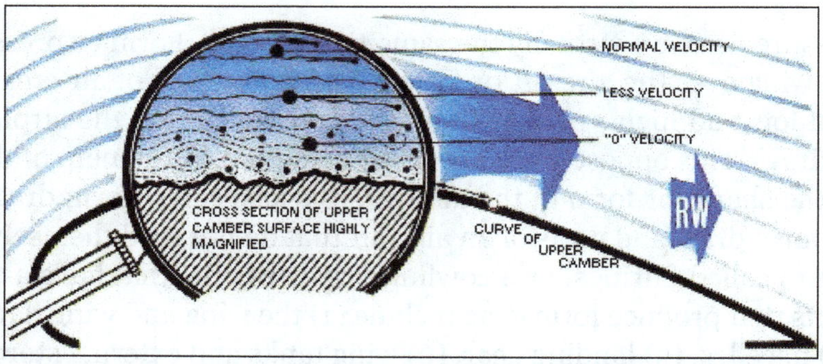

The leading edge of a wing will always produce a certain amount of friction drag.

If the airflow were ideal, that is, inviscid, the air would simply slip over the surface of a smooth plate with velocity V¥. At all points along the surface of the plate, the velocity distribution (that is, the variation in velocity as one move perpendicularly away from the surface) would be a uniform constant value of V∞. No drag would result if the airflow were frictionless (inviscid).

Under real conditions, however, a very thin film of air molecules adheres to the surface. This is the very important no-slip condition. It states that at the surface of a body, the airflow velocity is zero. As one moves away from the body, the velocity of the air gradually increases until, at some point, the velocity becomes a constant value; in the case of a flat plate this value is V¥. The layer of air where the velocity is changing from zero to a constant value is called the boundary layer. Within the boundary layer, there are relative velocities between the layers and an internal friction is present. This internal friction extends to the surface of the body. The cumulative effect of all these friction forces is to produce drag on the plate. This drag is referred to as skin-friction drag.

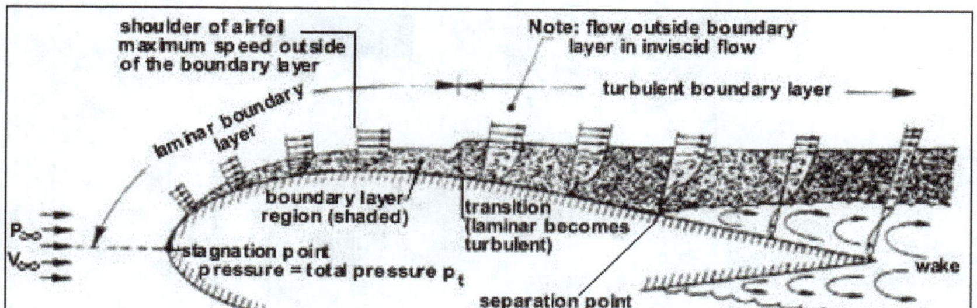

Real fluid flow about an airfoil. The thickness of the boundary layers and wake are greatly exaggerated. The bottom flow along lower surface is the same as on the upper surface.

Initially, near the leading edge of a flat, smooth plate, one has a laminar flow (the flow is layered) and the boundary layer also is steady and layered—hence, a laminar boundary layer. As one moves farther downstream, viscosity continues to act, and the laminar boundary layer thickens as more and more air is slowed down by internal friction. Eventually, a point is reached on the plate where the laminar boundary layer undergoes transition and becomes a turbulent boundary layer. As is usual for turbulent flow, there is random motion in the boundary layer as well as the downstream-directed motion. There is no slip at the surface of the plate. Another important difference from the laminar boundary layer is the fact that the velocity builds up more quickly as one move away from the wall, although the total boundary-layer thickness is greater. The turbulent boundary layer farther away from the wall reenergizes the slower moving air nearer the wall. This condition can be seen by comparing the profile of the laminar boundary layer with the profile of the turbulent boundary layer.

The Reynolds number has an important effect on the boundary layer. As the Reynolds number increases (caused by increasing the airflow speed and decreasing the viscosity), the boundary layer thickens more slowly. However, even though the Reynolds number becomes large, the velocity at the surface of the body must be zero. Thus, the boundary layer never disappears.

It is interesting to note that a typical thickness of the boundary layer on an aircraft wing is generally less than a centimetre (2.5 inches). Yet, the velocity must vary from zero at the surface of the wing to hundreds of meters per second at the outer edge of the boundary layer. It is evident that tremendous shearing forces (internal friction) must be acting in this region. This gives rise to the skin-friction drag.

Applied to an airfoil in a real airflow, the same free-stream velocity V¥ and free-stream static pressure p¥ apply. The field of air ahead of the airfoil is only slightly modified and for all practical purposes, the velocities and static pressures are the same as for the ideal fluid case. Again a stagnation point (a point with no motion) occurs at the leading edge of the airfoil and the pressure reaches its maximum value of pt at this point (total or stagnation pressure). From this point on along the airfoil, the picture changes.

In then flat plate, a boundary layer begins to form because of viscosity. This boundary layer is very thin and outside of it, the flow acts very much like that of an ideal fluid. Also, the static pressure acting on the surface of the airfoil is determined by the static pressure outside the boundary layer. This pressure is transmitted through the boundary layer to the surface and thus acts as if the boundary layer were not present at all. But the boundary layer feels this static pressure and will respond to it.

Over the front surface of the airfoil up to the shoulder, an assisting favourable pressure gradient exists (pressure decreasing with distance downstream). The airflow speeds up along the airfoil. The flow is laminar and a laminar boundary layer is present. This laminar boundary layer grows in thickness along the airfoil. When the shoulder is reached, however, the air molecules are moving slower than in the ideal fluid case. This is an unfavourable condition because the previous ideal flow just came to rest at the trailing edge of the airfoil. It would appear now, with viscosity present, that the flow will come to rest at some distance before the trailing edge is reached.

As the airflow moves from the shoulder to the rear surface of the airfoil, the static-pressure gradient is unfavourable (increasing pressure with downstream distance). The air molecules must push against both this unfavourable pressure gradient and the viscous forces. At the transition point, the character of the airflow changes and the laminar boundary layer quickly becomes a turbulent boundary layer. This turbulent boundary layer continues to thicken downstream.

Pushing against an unfavourable pressure gradient and viscosity is too much for the airflow, and at some point, the airflow stops completely. The boundary layer has stalled short of reaching the trailing edge. (Remember that the airflow reached the trailing edge before stopping in the ideal fluid case).

This stall point is known as the separation point. All along a line starting from this point outward into the airflow, the airflow is stalling. Beyond this line, the airflow is actually moving backward, upstream toward the nose before turning around. This is a region of eddies and whirlpools and represents "dead" air that is disrupting the flow field away from the airfoil. Thus, the airflow outside the dead air region is forced to flow away and around it. The region of eddies is called the wake behind the airfoil.

Up to the separation point, the difference between the static-pressure distribution for ideal fluid flow and real airflow is not very large but once separation occurs, the pressure field in greatly modified. In the ideal fluid case, the net static-pressure force acting on the front surface of the airfoil (up to the shoulder) parallel to the free stream exactly opposed and cancelled that acting on the rear surfaces of the airfoil. Under real airflow conditions, however, this symmetry and cancellation of forces is destroyed. The net static-pressure force acting on the front surface parallel to the free-stream direction now exceeds that acting on the rear surface. The net result is a drag force due to the asymmetric pressure distribution called pressure drag. This is a drag in addition to the skin-friction drag due to the shearing forces (internal friction) in the boundary layer. Additionally, the modification of the static-pressure distribution causes a decrease in the pressure lift from the ideal fluid case. The effect of viscosity is that the lift is reduced and a total drag composed of skin-friction drag and pressure drag is present. Both of these are detrimental effects.

It should be emphasized that similar processes are occurring on all the components of the aircraft to one degree or another, not only the airfoil.

Thus, the effects of a real fluid flow are the result of the viscosity of the fluid. The viscosity causes a boundary layer and, hence, a skin-friction drags. The flow field is disrupted because of viscosity to the extent that a pressure drag arises. Also, the net pressure lift is reduced.

Interference Drag

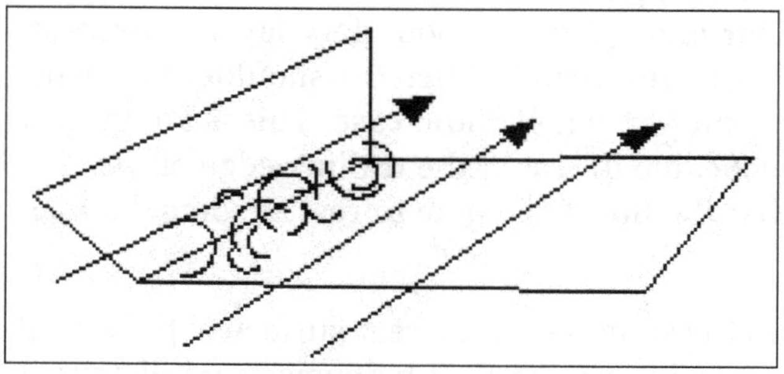

Interference Drag C.

Surfaces at angles to each other as in (C) create turbulence in the region of the joint. This occurs most frequently at the intersection of the fuselage and wing.

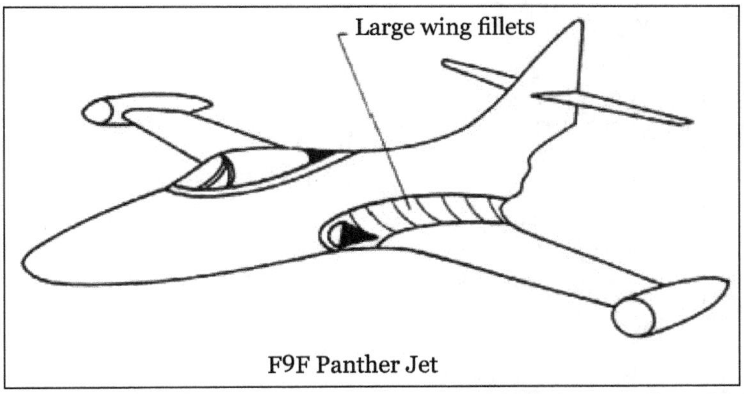

Grumman F9F Panther Jet with a large degree of filleting to reduce drag.

Lift versus Drag

An aircraft with a given total gross weight can be operated in level flight over a range of power settings and airspeeds. Since Lift and Weight must be equal in order to maintain level flight, it is obvious that there is a relationship between Lift (L), Airspeed (V), and Angle of Attack (AT). This relationship can be "generalized" with the following expression.

Lift = Angle of Attack × Velocity

Since angle of attack and speed also have a relationship to Induced Drag and Parasite Drag, the relationship of Lift/Drag is shown by the graph below:

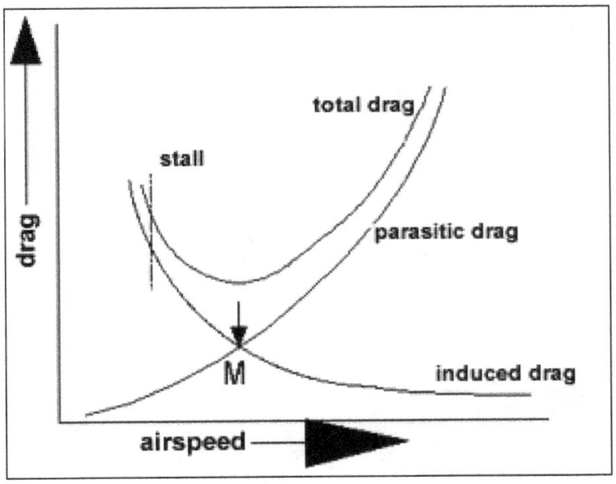

Parasite drag increases with speed. Induced drag decreases with speed. The sum of the two drags (Total Drag curve) shows that there is only one airspeed for a given airplane and load that provides minimum total drag. This is the point M which is the maximum lift over drag ratio (L/D). It is the airspeed at which the aircraft can glide the farthest without power (maximum glide range). This is the airspeed which should immediately be set up in the event of a power failure. This maximum glide airspeed is different for each aircraft design. The Pilot Operating Handbook should be consulted for this airspeed and the pilot should memorize it to eliminate need to search manuals during an emergency.

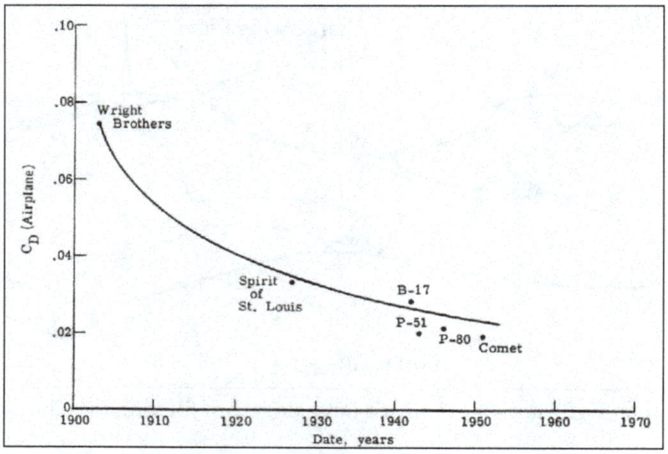

Decrease in airplane drag coefficient with time.

Lift on Aircraft

The pilot can control the lift. Any time the control yoke or stick is moved fore or aft, the AOA is changed. As the AOA increases, lift increases (all other factors being equal). When the aircraft reaches the maximum AOA, lift begins to diminish rapidly. This is the stalling AOA, known as CL-MAX critical AOA. Examine figure, noting how the CL increases until the critical AOA is reached, then decreases rapidly with any further increase in the AOA.

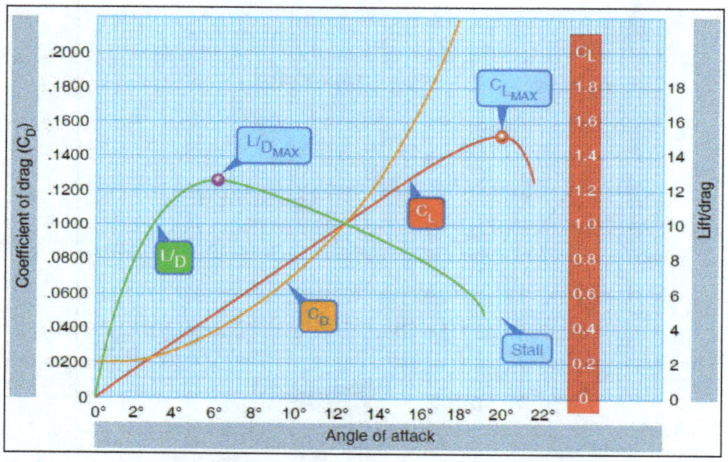

Coefficients of lift and drag at various angles of attack.

The shape of the wing or rotor cannot be effective unless it continually keeps "attacking" new air. If an aircraft is to keep flying, the lift-producing airfoil must keep moving. In a helicopter or gyroplane, this is accomplished by the rotation of the rotor blades. For other types of aircraft, such as airplanes, weight shift control, or gliders, air must be moving across the lifting surface. This is accomplished by the forward speed of the aircraft. Lift is proportional to the square of the aircraft's velocity. For example, an airplane traveling at 200 knots has four times the lift as the same airplane traveling at 100 knots, if the AOA and other factors remain constant.

$$L = \frac{C_L \cdot \rho \cdot V^2 \cdot S}{2}$$

The lift equation exemplifies this mathematically and supports that doubling of the airspeed will result in four times the lift. As a result, one can see that velocity is an important component to the production of lift, which itself can be affected through varying AOA. When examining the equation, lift (L) is determined through the relationship of the air density (ρ), the airfoil velocity (V), the surface area of the wing (S) and the coefficient of lift (CL) for a given airfoil. Taking the equation further, one can see an aircraft could not continue to travel in level flight at a constant altitude and maintain the same AOA if the velocity is increased. The lift would increase and the aircraft would climb as a result of the increased lift force or speed up. Therefore, to keep the aircraft straight and level (not accelerating upward) and in a state of equilibrium, as velocity is increased, lift must be kept constant. This is normally accomplished by reducing the AOA by lowering the nose. Conversely, as the aircraft is slowed, the decreasing velocity requires increasing the AOA to maintain lift sufficient to maintain flight. There is, of course, a limit to how far the AOA can be increased, if a stall is to be avoided.

All other factors being constant, for every AOA there is a corresponding airspeed required to maintain altitude in steady, unaccelerated flight (true only if maintaining level flight). Since an airfoil always stalls at the same AOA, if increasing weight, lift must also be increased. The only method of increasing lift is by increasing velocity if the AOA is held constant just short of the "critical," or stalling, AOA (assuming no flaps or other high lift devices).

Lift and drag also vary directly with the density of the air. Density is affected by several factors: pressure, temperature, and humidity. At an altitude of 18,000 feet, the density of the air has one-half the density of air at sea level. In order to maintain its lift at a higher altitude, an aircraft must fly at a greater true airspeed for any given AOA.

Warm air is less dense than cool air, and moist air is less dense than dry air. Thus, on a hot humid day, an aircraft must be flown at a greater true airspeed for any given AOA than on a cool, dry day.

If the density factor is decreased and the total lift must equal the total weight to remain in flight, it follows that one of the other factors must be increased. The factor usually increased is the airspeed or the AOA because these are controlled directly by the pilot.

Lift varies directly with the wing area, provided there is no change in the wing's planform. If the wings have the same proportion and airfoil sections, a wing with a planform area of 200 square feet lifts twice as much at the same AOA as a wing with an area of 100 square feet.

Two major aerodynamic factors from the pilot's viewpoint are lift and airspeed because they can be controlled readily and accurately. Of course, the pilot can also control density by adjusting the altitude and can control wing area if the aircraft happens to have flaps of the type that enlarge wing area. However, for most situations, the pilot controls lift and airspeed to maneuver an aircraft. For instance, in straight-and-level flight, cruising along at a constant altitude, altitude is maintained by adjusting lift to match the aircraft's velocity or cruise airspeed, while maintaining a state of equilibrium in which lift equals weight. In an approach to landing, when the pilot wishes to land as slowly as practical, it is necessary to increase AOA near maximum to maintain lift equal to the weight of the aircraft.

Lift/Drag Ratio

The lift-to-drag ratio (L/D) is the amount of lift generated by a wing or airfoil compared to its drag. A ratio of L/D indicates airfoil efficiency. Aircraft with higher L/D ratios are more efficient than those with lower L/D ratios. In unaccelerated flight with the lift and drag data steady, the proportions of the coefficient of lift (CL) and coefficient of drag (CD) can be calculated for specific AOA.

The coefficient of lift is dimensionless and relates the lift generated by a lifting body, the dynamic pressure of the fluid flow around the body, and a reference area associated with the body. The coefficient of drag is also dimensionless and is used to quantify the drag of an object in a fluid environment, such as air, and is always associated with a particular surface area.

The L/D ratio is determined by dividing the C_L by the C_D, which is the same as dividing the lift equation by the drag equation as all of the variables, aside from the coefficients, cancel out. The lift and drag equations are as follows (L = Lift in pounds; D = Drag; C_L = coefficient of lift; ρ = density (expressed in slugs per cubic feet); V = velocity (in feet per second); q = dynamic pressure per square foot ($q = 1/2\ \rho v^2$); S = the area of the lifting body (in square feet); and C_D = Ratio of drag pressure to dynamic pressure):

$$D = \frac{C_D \cdot \rho \cdot V^2 \cdot S}{2}$$

Typically at low AOA, the coefficient of drag is low and small changes in AOA create only slight changes in the coefficient of drag. At high AOA, small changes in the AOA cause significant changes in drag. The shape of an airfoil, as well as changes in the AOA, affects the production of lift.

In figure, that the coefficient of lift curve (red) reaches its maximum for this particular wing section at 20° AOA and then rapidly decreases. 20° AOA is therefore the critical angle of attack. The coefficient of drag curve (orange) increases very rapidly from 14° AOA and completely overcomes the lift curve at 21° AOA. The lift/drag ratio (green) reaches its maximum at 6° AOA, meaning that at this angle, the most lift is obtained for the least amount of drag.

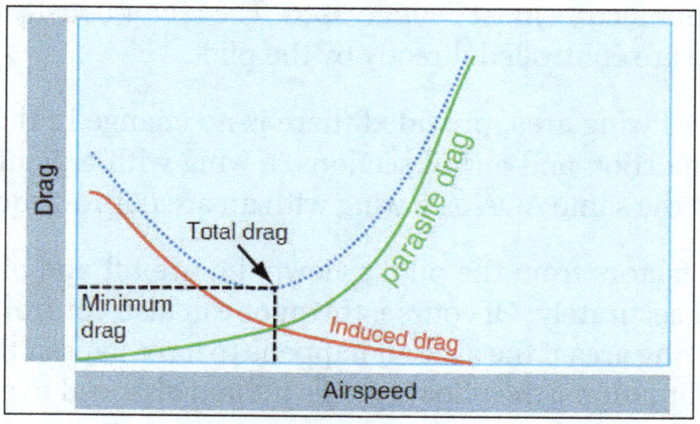

Drag versus speed.

The maximum lift/drag ratio (L/D_{MAX}) occurs at one specific C_L and AOA. If the aircraft is operated in steady flight at L/D_{MAX}, the total drag is at a minimum. Any AOA lower or higher than that for L/D_{MAX} reduces the L/D and consequently increases the total drag for a given aircraft's lift. Figure depicts the L/D_{MAX} by the lowest portion of the blue line labeled "total drag." The configuration of an aircraft has a great effect on the L/D.

Weight Force on Aircraft

Gravity is the pulling force that tends to draw all bodies to the center of the earth. The center of gravity may be considered as a point at which all the weight of the aircraft is concentrated. If the aircraft were supported at its exact center of gravity, it would balance in any attitude. It will be noted that center of gravity is of major importance in an aircraft, for its position has a great bearing upon stability. The allowable location of the center of gravity is determined by the general design of each particular aircraft. The designers determine how far the center of pressure (CP) will travel. It is important to understand that an aircraft's weight is concentrated at the center of gravity and the aerodynamic forces of lift occur at the CP. When the center of gravity is forward of the CP, there is a natural tendency for the aircraft to want to pitch nose down. If the CP is forward of the center of gravity, a nose up pitching moment is created. Therefore, designers fix the aft limit of the center of gravity forward of the CP for the corresponding flight speed in order to retain flight equilibrium.

Weight has a definite relationship to lift. This relationship is simple, but important in understanding the aerodynamics of flying. Lift is the upward force on the wing acting perpendicular to the relative wind and perpendicular to the aircraft's lateral axis. Lift is required to counteract the aircraft's weight. In stabilized level flight, when the lift force is equal to the weight force, the aircraft is in a state of equilibrium and neither accelerates upward or downward. If lift becomes less than weight, the vertical speed will decrease. When lift is greater than weight, the vertical speed will increase.

Wingtip Vortices

Formation of Vortices

The action of the airfoil that gives an aircraft lift also causes induced drag. When an airfoil is flown at a positive AOA, a pressure differential exists between the upper and lower surfaces of the airfoil. The pressure above the wing is less than atmospheric pressure and the pressure below the wing is equal to or greater than atmospheric pressure. Since air always moves from high pressure toward low pressure, and the path of least resistance is toward the airfoil's tips, there is a spanwise movement of air from the bottom of the airfoil outward from the fuselage around the tips. This flow of air results in "spillage" over the tips, thereby setting up a whirlpool of air called a vortex.

At the same time, the air on the upper surface has a tendency to flow in toward the fuselage and off the trailing edge. This air current forms a similar vortex at the inboard portion of the trailing edge of the airfoil, but because the fuselage limits the inward flow, the vortex is insignificant. Consequently, the deviation in flow direction is greatest at the outer tips where the unrestricted lateral flow is the strongest.

As the air curls upward around the tip, it combines with the downwash to form a fast-spinning trailing vortex. These vortices increase drag because of energy spent in producing the turbulence. Whenever an airfoil is producing lift, induced drag occurs and wingtip vortices are created.

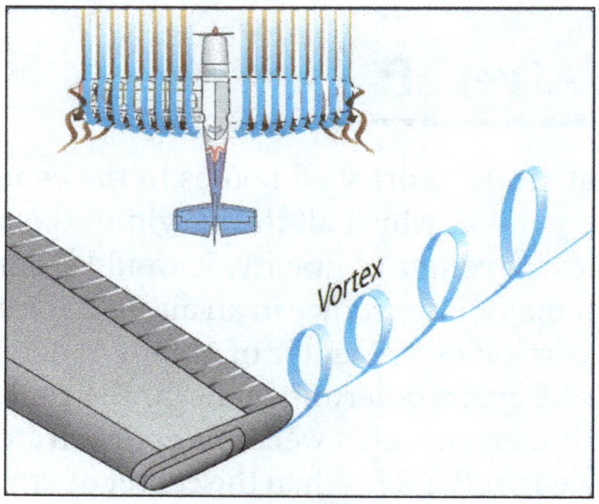

Wingtip vortices.

Just as lift increases with an increase in AOA, induced drag also increases. This occurs because as the AOA is increased, there is a greater pressure difference between the top and bottom of the airfoil, and a greater lateral flow of air; consequently, this causes more violent vortices to be set up, resulting in more turbulence and more induced drag.

In figure, it is easy to see the formation of wingtip vortices. The intensity or strength of the vortices is directly proportional to the weight of the aircraft and inversely proportional to the wingspan and speed of the aircraft. The heavier and slower the aircraft, the greater the AOA and the stronger the wingtip vortices. Thus, an aircraft will create wingtip vortices with maximum strength occurring during the takeoff, climb, and landing phases of flight. These vortices lead to a particularly dangerous hazard to flight, wake turbulence.

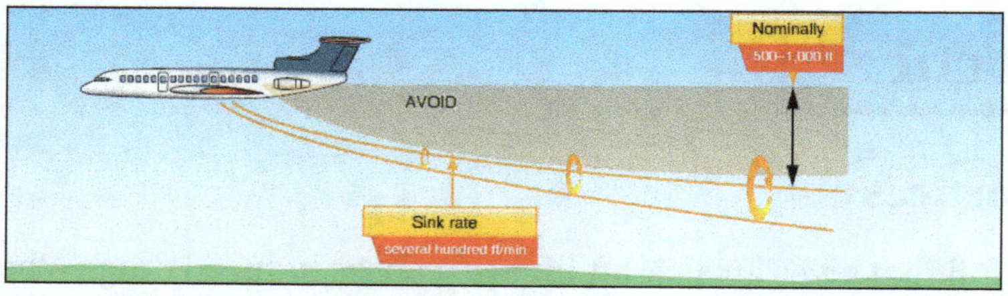

Avoid following another aircraft at an altitude within 1,000 feet.

Avoiding Wake Turbulence

Wingtip vortices are greatest when the generating aircraft is “heavy, clean, and slow.” This condition is most commonly encountered during approaches or departures because an aircraft’s AOA are at the highest to produce the lift necessary to land or take off. To minimize the chances of flying through an aircraft’s wake turbulence:

- Avoid flying through another aircraft’s flight path.
- Rotate prior to the point at which the preceding aircraft rotated when taking off behind another aircraft.

- Avoid following another aircraft on a similar flight path at an altitude within 1,000 feet.
- Approach the runway above a preceding aircraft's path when landing behind another aircraft and touch down after the point at which the other aircraft wheels contacted the runway.

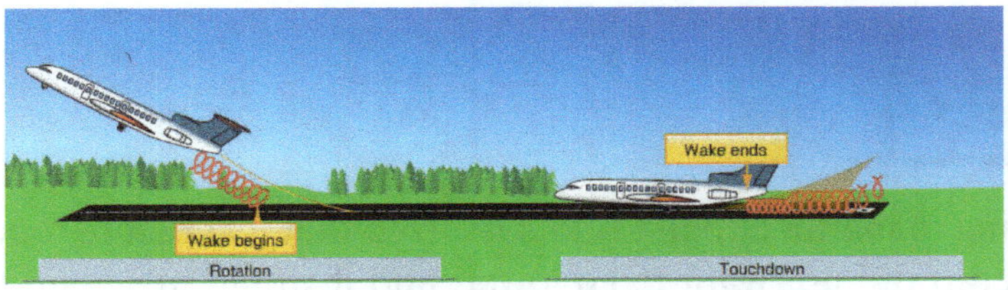

Avoid turbulence from another aircraft.

A hovering helicopter generates a down wash from its main rotor(s) similar to the vortices of an airplane. Pilots of small aircraft should avoid a hovering helicopter by at least three rotor disc diameters to avoid the effects of this down wash. In forward flight, this energy is transformed into a pair of strong, high-speed trailing vortices similar to wing-tip vortices of larger fixed-wing aircraft. Helicopter vortices should be avoided because helicopter forward flight airspeeds are often very slow and can generate exceptionally strong wake turbulence.

Wind is an important factor in avoiding wake turbulence because wingtip vortices drift with the wind at the speed of the wind. For example, a wind speed of 10 knots causes the vortices to drift at about 1,000 feet in a minute in the wind direction. When following another aircraft, a pilot should consider wind speed and direction when selecting an intended takeoff or landing point. If a pilot is unsure of the other aircraft's takeoff or landing point, approximately 3 minutes provides a margin of safety that allows wake turbulence dissipation.

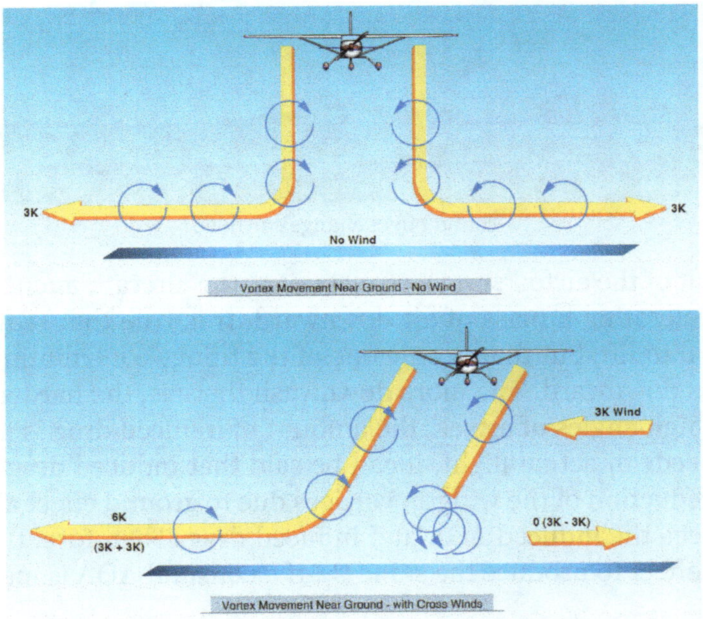

Figure shows when the vortices of larger aircraft sink close to the ground (within 100 to 200 feet), they tend to move laterally over the ground at a speed of 2 or 3 knots (top). A crosswind will decrease the lateral movement of the upwind vortex and increase the movement of the downwind vortex. Thus a light wind with a cross runway component of 1 to 5 knots could result in the upwind vortex remaining in the touchdown zone for a period of time and hasten the drift of the downwind vortex toward another runway (bottom).

Ground Effect

Ever since the beginning of manned flight, pilots realized that just before touchdown it would suddenly feel like the aircraft did not want to go lower, and it would just want to go on and on. This is due to the air that is trapped between the wing and the landing surface, as if there were an air cushion. This phenomenon is called ground effect.

When an aircraft in flight comes within several feet of the surface, ground or water, a change occurs in the three dimensional flow pattern around the aircraft because the vertical component of the airflow around the wing is restricted by the surface. This alters the wing's upwash, downwash, and wingtip vortices. Ground effect, then, is due to the interference of the ground (or water) surface with the airflow patterns about the aircraft in flight. While the aerodynamic characteristics of the tail surfaces and the fuselage are altered by ground effect, the principal effects due to proximity of the ground are the changes in the aerodynamic characteristics of the wing. As the wing encounters ground effect and is maintained at a constant AOA, there is consequent reduction in the upwash, downwash, and wingtip vortices.

Ground effect changes airflow.

Induced drag is a result of the airfoil's work of sustaining the aircraft, and a wing or rotor lifts the aircraft simply by accelerating a mass of air downward. It is true that reduced pressure on top of an airfoil is essential to lift, but that is only one of the things contributing to the overall effect of pushing an air mass downward. The more downwash there is, the harder the wing pushes the mass of air down. At high angles of attack, the amount of induced drag is high; since this corresponds to lower airspeeds in actual flight, it can be said that induced drag predominates at low speed. However, the reduction of the wingtip vortices due to ground effect alters the spanwise lift distribution and reduces the induced AOA and induced drag. Therefore, the wing will require a lower AOA in ground effect to produce the same C_L. If a constant AOA is maintained, an increase in C_L results.

Ground effect also alters the thrust required versus velocity. Since induced drag predominates at low speeds, the reduction of induced drag due to ground effect will cause a significant reduction of thrust required (parasite plus induced drag) at low speeds. Due to the change in upwash, downwash, and wingtip vortices, there may be a change in position (installation) error of the airspeed system associated with ground effect. In the majority of cases, ground effect causes an increase in the local pressure at the static source and produces a lower indication of airspeed and altitude. Thus, an aircraft may be airborne at an indicated airspeed less than that normally required.

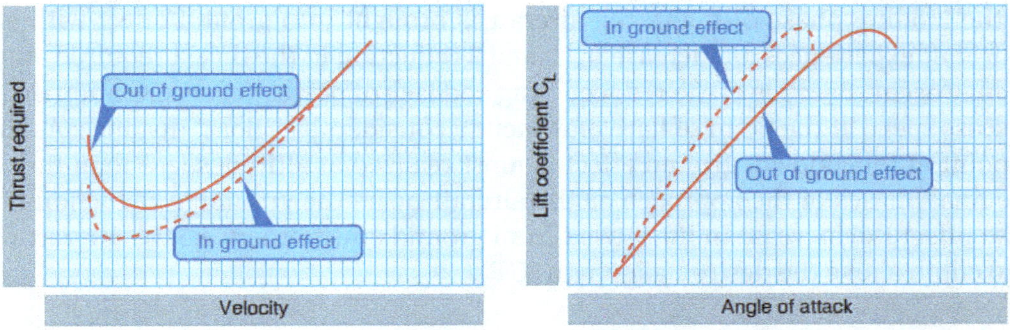

Ground effect changes drag and lift.

In order for ground effect to be of significant magnitude, the wing must be quite close to the ground. One of the direct results of ground effect is the variation of induced drag with wing height above the ground at a constant C_L. When the wing is at a height equal to its span, the reduction in induced drag is only 1.4 percent. However, when the wing is at a height equal to one-fourth its span, the reduction in induced drag is 23.5 percent and, when the wing is at a height equal to one-tenth its span, the reduction in induced drag is 47.6 percent. Thus, a large reduction in induced drag takes place only when the wing is very close to the ground. Because of this variation, ground effect is most usually recognized during the liftoff for takeoff or just prior to touchdown when landing.

During the takeoff phase of flight, ground effect produces some important relationships. An aircraft leaving ground effect after takeoff encounters just the reverse of an aircraft entering ground effect during landing. The aircraft leaving ground effect will:

- Require an increase in AOA to maintain the same CL.
- Experience an increase in induced drag and thrust required.
- Experience a decrease in stability and a nose-up change in moment.
- Experience a reduction in static source pressure and increase in indicated airspeed.

Ground effect must be considered during takeoffs and landings. For example, if a pilot fails to understand the relationship between the aircraft and ground effect during takeoff, a hazardous situation is possible because the recommended takeoff speed may not be achieved. Due to the reduced drag in ground effect, the aircraft may seem capable of takeoff well below the recommended speed. As the aircraft rises out of ground effect with a deficiency of speed, the greater induced drag may result in marginal initial climb performance. In extreme conditions, such as high gross weight, high density altitude, and high temperature, a deficiency of airspeed during takeoff may permit the aircraft to become airborne but be incapable of sustaining flight out of ground effect. In this case, the aircraft may become airborne initially with a deficiency of speed and then settle back to the runway.

A pilot should not attempt to force an aircraft to become airborne with a deficiency of speed. The manufacturer's recommended takeoff speed is necessary to provide adequate initial climb performance. It is also important that a definite climb be established before a pilot retracts the landing gear or flaps. Never retract the landing gear or flaps prior to establishing a positive rate of climb and only after achieving a safe altitude.

If, during the landing phase of flight, the aircraft is brought into ground effect with a constant AOA, the aircraft experiences an increase in C_L and a reduction in the thrust required, and a "floating" effect may occur. Because of the reduced drag and lack of power-off deceleration in ground effect, any excess speed at the point of flare may incur a considerable "float" distance. As the aircraft nears the point of touchdown, ground effect is most realized at altitudes less than the wingspan. During the final phases of the approach as the aircraft nears the ground, a reduction of power is necessary to offset the increase in lift caused from ground effect otherwise the aircraft will have a tendency to climb above the desired glidepath (GP).

The Three Axes of an Airplane

An airplane has three axes about/around which it can move in flight. In figure, you are seeing all three of these axes:

- Longitudinal axis: from the nose to the tail of the airplane.
- Lateral axis: from wingtip to wingtip.
- Vertical axis: from the bottom of the airplane to the top of the airplane.

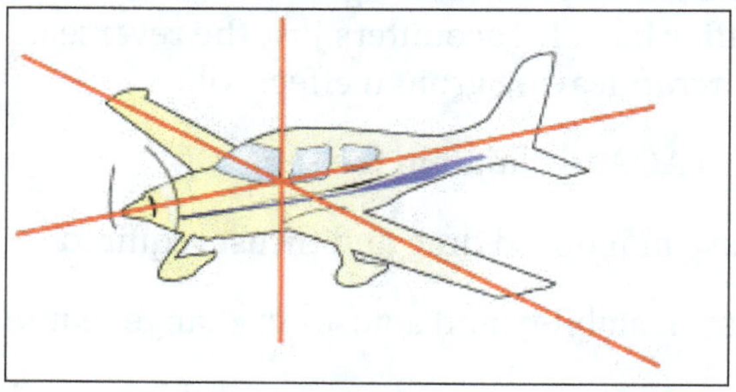

Each of these three axes meets at the center of gravity (center of gravity) of the plane. The center of gravity is the point at which the airplane would balance if you could lift it up by an imaginary string that attached at that exact point. You can think of the center of gravity also as a kind of fulcrum (like on a teeter-totter) that the plane rolls and pitches around. Anyway, back to the axes of the plane. Here's what you need to know about them and why they are important.

The Longitudinal axis (the one that runs the length of the plane from nose to tail) is the axis that stays fixed when the airplane "rolls" or "banks" – such as in a turn. In this case, the plane is rotating "about" or "around" the longitudinal axis. This is caused by the airplane's ailerons, which change the camber of the wing and increase its lift on one side, making the wing climb, and spoil the lift on the other side, making the wing drop.

The Lateral axis (the one that runs from wingtip to wingtip) stays fixed when the plane "pitches" – raising or lowering the nose (such as for a climb or a descent). The plane pitches about the lateral axis. This is done using the airplane's elevators. The elevators change the shape of the horizontal stabilizer, causing it to decrease lift (tail goes down, nose goes up) or increase lift (tail goes up, nose goes down). Some aircraft have "stabilators," where the entire horizontal surface moves instead of just the elevator, but the concept (and result) is still the same.

The last one is the Vertical axis, which runs vertically (up and down) through the fuselage. This one stays fixed when the airplane "yaws" – meaning the nose moves left or right. When the plane yaws, it is turning about the vertical axis. This is like turning a car (the car doesn't roll or pitch; it just turns, or "yaws"). This can be done by moving the rudder, which is the movable control surface on the vertical stabilizer (the upright portion of the tail). Moving the rudder right puts it into the airflow and pushes the tail to the left (and the nose to the right). Generally, the rudder is used in tandem with the ailerons to coordinate an airplane's turns, because when an airplane banks, there is a change in drag, making the nose want to move away from the turn initially. the rudder is used to "yaw" the nose the right way and keep the whole plane moving in the direction the pilot (you) want it to go.

Moment and Moment Arm

A study of physics shows that a body that is free to rotate will always turn about its center of gravity. In aerodynamic terms, the mathematical measure of an aircraft's tendency to rotate about its center of gravity is called a "moment." A moment is said to be equal to the product of the force applied and the distance at which the force is applied. (A moment arm is the distance from a datum to the applied force.) For aircraft weight and balance computations, "moments" are expressed in terms of the distance of the arm times the aircraft's weight, or simply, inch-pounds.

Aircraft designers locate the fore and aft position of the aircraft's center of gravity as nearly as possible to the 20 percent point of the mean aerodynamic chord (MAC). If the thrust line is designed to pass horizontally through the center of gravity, it will not cause the aircraft to pitch when power is changed, and there will be no difference in moment due to thrust for a power-on or power-off condition of flight. Although designers have some control over the location of the drag forces, they are not always able to make the resultant drag forces pass through the center of gravity of the aircraft. However, the one item over which they have the greatest control is the size and location of the tail. The objective is to make the moments (due to thrust, drag, and lift) as small as possible and, by proper location of the tail, to provide the means of balancing an aircraft longitudinally for any condition of flight.

The pilot has no direct control over the location of forces acting on the aircraft in flight, except for controlling the center of lift by changing the AOA. Such a change, however, immediately involves changes in other forces. Therefore, the pilot cannot independently change the location of one force without changing the effect of others. For example, a change in airspeed involves a change in lift, as well as a change in drag and a change in the up or down force on the tail. As forces such as turbulence and gusts act to displace the aircraft, the pilot reacts by providing opposing control forces to counteract this displacement.

Some aircraft are subject to changes in the location of the center of gravity with variations of load. Trimming devices are used to counteract the forces set up by fuel burnoff, and loading or off-loading of passengers or cargo. Elevator trim tabs and adjustable horizontal stabilizers comprise the most common devices provided to the pilot for trimming for load variations. Over the wide ranges of balance during flight in large aircraft, the force which the pilot has to exert on the controls would become excessive and fatiguing if means of trimming were not provided.

Stalls

An aircraft stall results from a rapid decrease in lift caused by the separation of airflow from the wing's surface brought on by exceeding the critical AOA. A stall can occur at any pitch attitude or airspeed. Stalls are one of the most misunderstood areas of aerodynamics because pilots often believe an airfoil stops producing lift when it stalls. In a stall, the wing does not totally stop producing lift. Rather, it cannot generate adequate lift to sustain level flight.

Since the C_L increases with an increase in AOA, at some point the C_L peaks and then begins to drop off. This peak is called the CL_{-MAX}. The amount of lift the wing produces drops dramatically after exceeding the CL_{-MAX} or critical AOA, but as stated above, it does not completely stop producing lift.

In most straight-wing aircraft, the wing is designed to stall the wing root first. The wing root reaches its critical AOA first making the stall progress outward toward the wingtip. By having the wing root stall first, aileron effectiveness is maintained at the wingtips, maintaining controllability of the aircraft. Various design methods are used to achieve the stalling of the wing root first. In one design, the wing is "twisted" to a higher AOA at the wing root. Installing stall strips on the first 20–25 percent of the wing's leading edge is another method to introduce a stall prematurely.

The wing never completely stops producing lift in a stalled condition. If it did, the aircraft would fall to the Earth. Most training aircraft are designed for the nose of the aircraft to drop during a stall, reducing the AOA and "unstalling" the wing. The nose-down tendency is due to the CL being aft of the CG. The CG range is very important when it comes to stall recovery characteristics. If an aircraft is allowed to be operated outside of the CG range, the pilot may have difficulty recovering from a stall. The most critical CG violation would occur when operating with a CG that exceeds the rear limit. In this situation, a pilot may not be able to generate sufficient force with the elevator to counteract the excess weight aft of the CG. Without the ability to decrease the AOA, the aircraft continues in a stalled condition until it contacts the ground.

The stalling speed of a particular aircraft is not a fixed value for all flight situations, but a given aircraft always stalls at the same AOA regardless of airspeed, weight, load factor, or density altitude. Each aircraft has a particular AOA where the airflow separates from the upper surface of the wing and the stall occurs. This critical AOA varies from approximately 16° to 20° depending on the aircraft's design. But each aircraft has only one specific AOA where the stall occurs.

There are three flight situations in which the critical AOA is most frequently exceeded: low speed, high speed, and turning.

One way the aircraft can be stalled in straight-and-level flight by flying too slowly. As the airspeed decreases, the AOA must be increased to retain the lift required for maintaining altitude. The lower the airspeed becomes, the more the AOA must be increased. Eventually, an AOA is reached that result in the wing not producing enough lift to support the aircraft, which then starts settling. If the airspeed is reduced further, the aircraft stalls because the AOA has exceeded the critical angle and the airflow over the wing is disrupted.

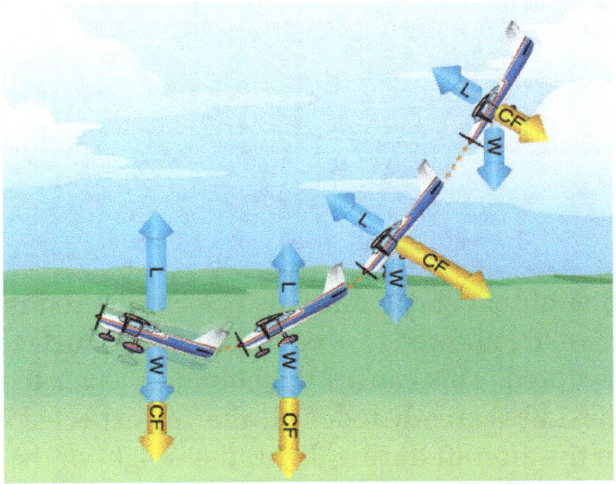

Forces exerted when pulling out of a dive.

Low speed is not necessary to produce a stall. The wing can be brought into an excessive AOA at any speed. For example, an aircraft is in a dive with airspeed of 100 knots when the pilot pulls back sharply on the elevator control. Gravity and centrifugal force prevent an immediate alteration of the flight path, but the aircraft's AOA changes abruptly from quite low to very high. Since the flight path of the aircraft in relation to the oncoming air determines the direction of the relative wind, the AOA is suddenly increased, and the aircraft would reach the stalling angle at a speed much greater than the normal stall speed.

The stalling speed of an aircraft is also higher in a level turn than in straight-and-level flight. Centrifugal force is added to the aircraft's weight and the wing must produce sufficient additional lift to counterbalance the load imposed by the combination of centrifugal force and weight. In a turn, the necessary additional lift is acquired by applying back pressure to the elevator control. This increases the wing's AOA and results in increased lift. The AOA must increase as the bank angle increases to counteract the increasing load caused by centrifugal force. If at any time during a turn the AOA becomes excessive, the aircraft stalls.

At this point, the action of the aircraft during a stall should be examined. To balance the aircraft aerodynamically, the CL is normally located aft of the CG. Although this makes the aircraft inherently nose-heavy, downwash on the horizontal stabilizer counteracts this condition. At the point of stall, when the upward force of the wing's lift diminishes below that required for sustained flight and the downward tail force decreases to a point of ineffectiveness, or causes it to have an upward force, an unbalanced condition exists. This causes the aircraft to pitch down abruptly, rotating about its CG. During this nose-down attitude, the AOA decreases and the airspeed again increase. The smooth flow of air over the wing begins again, lift returns, and the aircraft begins to fly again. Considerable altitude may be lost before this cycle is complete.

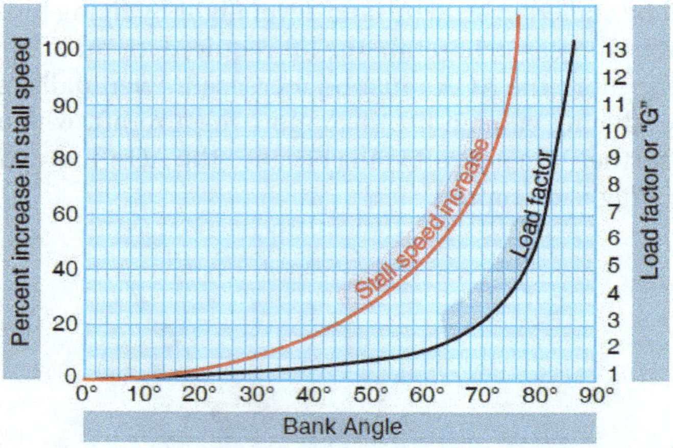

Increase in stall speed and load factor.

Airfoil shape and degradation of that shape must also be considered in a discussion of stalls. For example, if ice, snow, and frost are allowed to accumulate on the surface of an aircraft, the smooth airflow over the wing is disrupted. This causes the boundary layer to separate at an AOA lower than that of the critical angle. Lift is greatly reduced, altering expected aircraft performance. If ice is allowed to accumulate on the aircraft during flight, the weight of the aircraft is increased while the ability to generate lift is decreased. As little as 0.8 millimeter of ice on the upper wing surface increases drag and reduces aircraft lift by 25 percent.

Inflight ice formation.

Pilots can encounter icing in any season, anywhere in the country, at altitudes of up to 18,000 feet and sometimes higher. Small aircraft, including commuter planes, are most vulnerable because they fly at lower altitudes where ice is more prevalent. They also lack mechanisms common on jet aircraft that prevent ice buildup by heating the front edges of wings.

Icing can occur in clouds any time the temperature drops below freezing and super-cooled droplets build up on an aircraft and freeze. (Super-cooled droplets are still liquid even though the temperature is below 32 °Fahrenheit (F), or 0 °Celsius (C).

References

- What-is-aerodynamics, simwiki, docs: simscale.com, Retrieved 18 April, 2019
- Effect-of-wing-planform: aircraftsystemstech.com, Retrieved 7 May, 2019

- Angle-of-attack, theory: aviation-history.com, Retrieved 21 July, 2019
- Forces-acting-on-the-aircraft-thrust: flightliteracy.com, Retrieved 13 July, 2019
- Drag, aero, flight-training, training: pilotfriend.com, Retrieved 24 March , 2019
- Forces-acting-on-aircraft: aircraftsystemstech.com, Retrieved 1 June, 2019
- Ground-effect: flightliteracy.com, Retrieved 21 March , 2019
- Three-axes-of-an-airplane: smartflighttraining.com, Retrieved 12 April, 2019
- Moment-and-moment-arm: flightlearnings.com, Retrieved 23 June, 2019

4

Propulsion System in Aerospace Engineering

The system which provides the energy for a vehicle to move through outer space or air is termed as a propulsion system. Some of the areas of study related to the propulsion system are the propeller and the aircraft propeller vibration. This chapter has been carefully written to provide an easy understanding of the varied facets of propulsion systems related to aircraft as well as rocket propellants.

Propeller

An airplane moving through the air creates a drag force opposing its forward motion. If an airplane is to fly on a level path, there must be a force applied to it that is equal to the drag but acting forward. This force is called thrust. The work done by thrust is equal to the thrust times the distance it moves the airplane.

Work = Thrust × Distance

The power expended by thrust is equal to the thrust times the velocity at which it moves the airplane.

Power = Thrust × Velocity

If the power is measured in horsepower units, the power expended by the thrust is termed thrust horsepower.

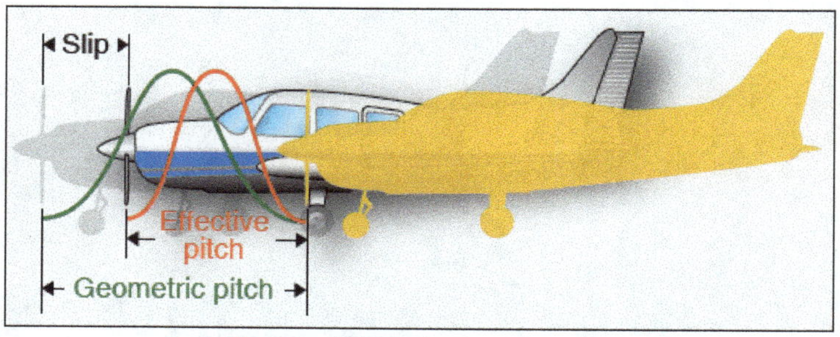

Effective pitch and geometric pitch.

The engine supplies brake horsepower through a rotating shaft, and the propeller converts it into thrust horsepower. In this conversion, some power is wasted. For maximum efficiency, the propeller must be designed to keep this waste as small as possible. Since the efficiency of any machine is the ratio of the useful power output to the power input, propeller efficiency is the ratio of thrust horsepower to brake horsepower. The usual symbol for propeller efficiency is the Greek letter η (eta). Propeller efficiency varies from 50 percent to 87 percent, depending on how much the propeller slips.

Pitch is not the same as blade angle, but because pitch is largely determined by blade angle, the two terms are often used interchangeably. An increase or decrease in one is usually associated with an increase or decrease in the other. Propeller slip is the difference between the geometric pitch of the propeller and its effective pitch. Geometric pitch is the distance a propeller should advance in one revolution with no slippage; effective pitch is the distance it actually advances. Thus, geometric or theoretical pitch is based on no slippage. Actual, or effective, pitch recognizes propeller slippage in the air. The relationship can be shown as:

Geometric pitch – Effective pitch = slip

Geometric pitch is usually expressed in pitch inches and calculated by using the following formula:

GP = 2 × π R × tangent of blade angle at 75 percent station

Where,

R = Radius at the 75 percent blade station

$\pi = 3.14$

Although blade angle and propeller pitch are closely related, blade angle is the angle between the face or chord of a blade section and the plane in which the propeller rotates. Blade angle, usually measured in degrees, is the angle between the chordline of the blade and the plane of rotation. The chordline of the propeller blade is determined in about the same manner as the chordline of an airfoil. In fact, a propeller blade can be considered as being composed of an infinite number of thin blade elements, each of which is a miniature airfoil section whose chord is the width of the propeller blade at that section. Because most propellers have a flat blade face, the chord line is often drawn along the face of the propeller blade.

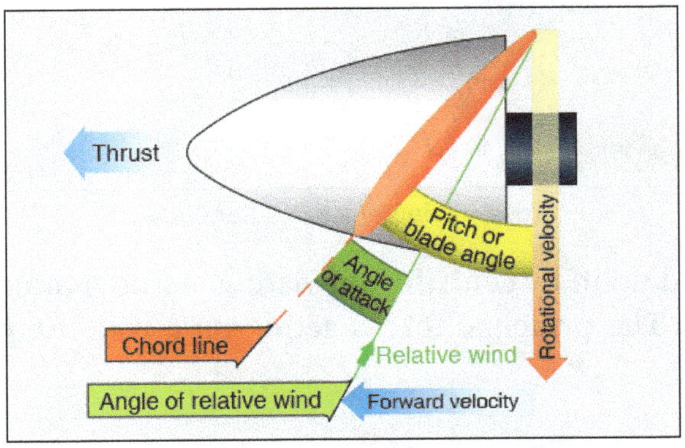

Propeller aerodynamic factors.

The typical propeller blade can be described as a twisted airfoil of irregular planform. Two views of a propeller blade are shown in figure. For purposes of analysis, a blade can be divided into segments that are located by station numbers in inches from the center of the blade hub. The cross-sections of each 6-inch blade segment are shown as airfoils in the right side of figure. Also identified in figure are the blade shank and the blade butt. The blade shank is the thick, rounded portion of the propeller blade near the hub and is designed to give strength to the blade. The blade butt, also called the blade base or root, is the end of the blade that fits in the propeller hub. The blade tip is that part of the propeller blade farthest from the hub, generally defined as the last 6 inches of the blade.

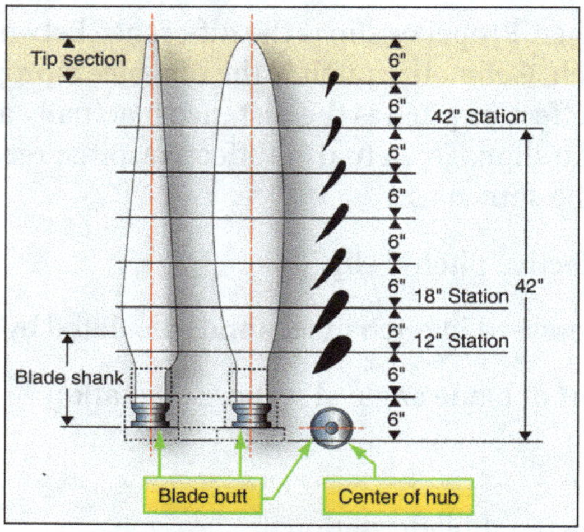

Typical propeller blade elements.

A cross-section of a typical propeller blade is shown in figure. This section or blade element is an airfoil comparable to a cross-section of an aircraft wing. The blade back is the cambered or curved side of the blade, similar to the upper surface of an aircraft wing. The blade face is the flat side of the propeller blade. The chord line is an imaginary line drawn through the blade from the leading edge to the trailing edge. The leading edge is the thick edge of the blade that meets the air as the propeller rotates.

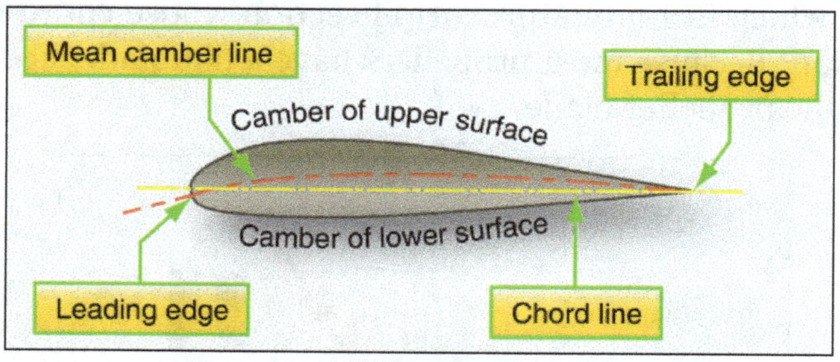

Cross-section of a propeller blade.

A rotating propeller is acted upon by centrifugal twisting, aerodynamic twisting, torque bending, and thrust bending forces. The principal forces acting on a rotating propeller are illustrated in figure.

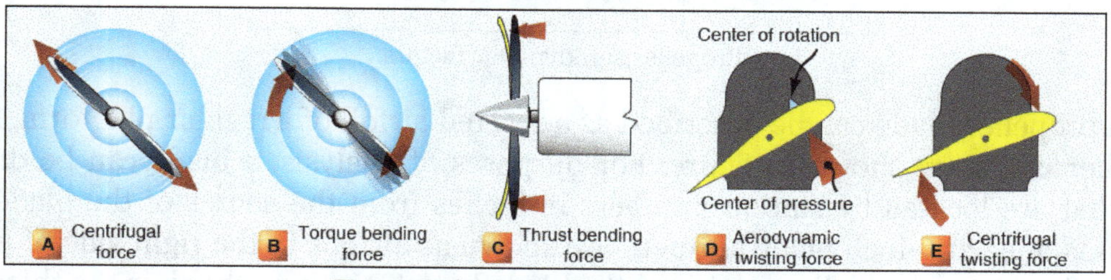

Forces acting on a rotating propeller.

Centrifugal force is a physical force that tends to throw the rotating propeller blades away from the hub. This is the most dominant force on the propeller. Torque bending force, in the form of air resistance, tends to bend the propeller blades in the direction opposite that of rotation. Thrust bending force is the thrust load that tends to bend propeller blades forward as the aircraft is pulled through the air. Aerodynamic twisting force tends to turn the blades to a high blade angle. Centrifugal twisting force, being greater than the aerodynamic twisting force, tends to force the blades toward a low blade angle.

At least two of these forces acting on the propellers blades are used to move the blades on a controllable pitch propeller. Centrifugal twisting force is sometimes used to move the blades to the low pitch position, while aerodynamic twisting force is used to move the blades into high pitch. These forces can be the primary or secondary forces that move the blades to the new pitch position.

A propeller must be capable of withstanding severe stresses, which are greater near the hub, caused by centrifugal force and thrust. The stresses increase in proportion to the rpm. The blade face is also subjected to tension from the centrifugal force and additional tension from the bending. For these reasons, nicks or scratches on the blade may cause very serious consequences. These could lead to cracks and failure of the blade.

A propeller must also be rigid enough to prevent fluttering, a type of vibration in which the ends of the blade twist back and forth at high frequency around an axis perpendicular to the engine crankshaft. Fluttering is accompanied by a distinctive noise, often mistaken for exhaust noise. The constant vibration tends to weaken the blade and eventually causes failure.

Aerodynamic Factors

To understand the action of a propeller, consider first its motion, which is both rotational and forward. Thus, as shown by the vectors of propeller forces in figure, a section of a propeller blade moves downward and forward. As far as the forces are concerned, the result is the same as if the blade were stationary and the air coming at it from a direction opposite its path. The angle at which this air (relative wind) strikes the propeller blade is called angle of attack (AOA). The air deflection produced by this angle causes the dynamic pressure at the engine side of the propeller blade to be greater than atmospheric pressure, creating thrust.

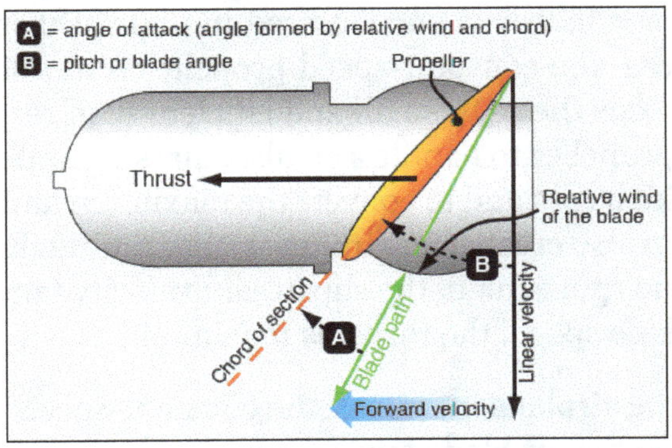

Propeller forces.

The shape of the blade also creates thrust because it is shaped like a wing. As the air flows past the propeller, the pressure on one side is less than that on the other. As in a wing, this difference in pressure produces a reaction force in the direction of the lesser pressure. The area above a wing has less pressure, and the force (lift) is upward. The area of decreased pressure is in front of a propeller which is mounted in a vertical instead of a horizontal position, and the force (thrust) is in a forward direction. Aerodynamically, thrust is the result of the propeller shape and the AOA of the blade.

Another way to consider thrust is in terms of the mass of air handled. In these terms, thrust is equal to the mass of air handled multiplied by the slipstream velocity minus the velocity of the airplane. Thus, the power expended in producing thrust depends on the mass of air moved per second. On the average, thrust constitutes approximately 80 percent of the torque (total horsepower absorbed by the propeller). The other 20 percent is lost in friction and slippage. For any speed of rotation, the horsepower absorbed by the propeller balances the horsepower delivered by the engine. For any single revolution of the propeller, the amount of air displaced (moved) depends on the blade angle, which determines the quantity or amount of mass of air the propeller moves. Thus, the blade angle is an excellent means of adjusting the load on the propeller to control the engine rpm. If the blade angle is increased, more loads is placed on the engine, tending to slow it down unless more power is applied. As an airfoil is moved through the air, it produces two forces: lift and drag. Increasing propeller blade angle increases the AOA and produces more lift and drag; this action increases the horsepower required to turn the propeller at a given rpm. Since the engine is still producing the same horsepower, the propeller slows down. If the blade angle is decreased, the propeller speeds up. Thus, the engine rpm can be controlled by increasing or decreasing the blade angle.

The blade angle is also an excellent method of adjusting the AOA of the propeller. On constant-speed propellers, the blade angle must be adjusted to provide the most efficient AOA at all engine and airplane speeds. Lift versus drag curves, which are drawn for propellers as well as wings, indicate that the most efficient AOA is a small one varying from 2° to 4° positive. The actual blade angle necessary to maintain this small AOA varies with the forward speed of the airplane. This is due to a change in the relative wind direction, which varies with aircraft speed.

Fixed-pitch and ground-adjustable propellers are designed for best efficiency at one rotation and forward speed. In other words, they are designed to fit a given airplane and engine combination. A propeller may be used that provides the maximum propeller efficiency for takeoff, climb, cruising, or high speeds. Any change in these conditions results in lowering the efficiency of both the propeller and the engine. A constant-speed propeller, however, keeps the blade angle adjusted for maximum efficiency for most conditions encountered in flight. During takeoff, when maximum power and thrust are required, the constant-speed propeller is at a low propeller blade angle or pitch. The low blade angle keeps the AOA small and efficient with respect to the relative wind. At the same time, it allows the propeller to handle a smaller mass of air per revolution. This light load allows the engine to turn at high rpm and to convert the maximum amount of fuel into heat energy in a given time. The high rpm also creates maximum thrust. Although the mass of air handled per revolution is small, the engine rpm is high, the slipstream velocity (air coming off the propeller) is high, and, with the low airplane speed, the thrust is maximum.

After liftoff, as the speed of the airplane increases, the constant-speed propeller changes to a higher angle (or pitch). Again, the higher blade angle keeps the AOA small and efficient with respect to the relative wind. The higher blade angle increases the mass of air handled per revolution. This decreases the engine rpm, reducing fuel consumption and engine wear, and keeps thrust at a maximum.

For climb after takeoff, the power output of the engine is reduced to climb power by decreasing the manifold pressure and increasing the blade angle to lower the rpm. Thus, the torque (horsepower absorbed by the propeller) is reduced to match the reduced power of the engine. The AOA is again kept small by the increase in blade angle. The greater mass of air handled per second, in this case, is more than offset by the lower slipstream velocity and the increase in airspeed.

At cruising altitude, when the airplane is in level flight and less power is required than is used in takeoff or climb, engine power is again reduced by lowering the manifold pressure and increasing the blade angle to decrease the rpm. Again, this reduces torque to match the reduced engine power; for, although the mass of air handled per revolution is greater, it is more than offset by a decrease in slipstream velocity and an increase in airspeed. The AOA is still small because the blade angle has been increased with an increase in airspeed. Pitch distribution is the twist in the blade from the shank to the blade tip, due to the variation in speeds that each section of the blade is traveling. The tip of the blade is traveling much faster than the inner portion of the blade.

Propeller Controls and Instruments

Turboprop propeller controls.

Fixed pitch propellers have no controls and require no adjustments in flight. The constant-speed propeller has a propeller control in the center pedestal between the throttle and the mixture control. The two positions for the control are increase rpm (full forward) and decrease rpm (pulled aft). This control is directly connected to the propeller governor and, by moving the control, adjusts the tension on the governor speeder spring. This control can also be used to feather the propeller in some aircraft by moving the control to the full decrease rpm position. The two main instruments used with the constant-speed propeller are the engine tachometer and the manifold pressure gauge. A rotation per minute (rpm) is controlled by the propeller control and the manifold pressure is adjusted by the throttle.

Constant Speed Aircraft Propeller

Hartzell Constant Speed: Nonfeathering

Hartzell propellers can be divided by Aluminum hub (compact) and steel hub. Hartzell compact aluminum propellers represent new concepts in basic design. They combine low weight and simplicity in design and rugged construction. In order to achieve these ends, the hub is made as compact as possible, utilizing aluminum alloy forgings for most of the parts. The hub shell is made in two halves, bolted together along the plane of rotation. This hub shell carries the pitch change mechanism and blade roots internally. The hydraulic cylinder, which provides power for changing the pitch, is mounted at the front of the hub. The propeller can be installed only on engines with flanged mounting provisions.

One model of nonfeathering aluminum hub constant-speed propeller utilizes oil pressure from a governor to move the blades into high pitch (reduced rpm). The centrifugal twisting moment of the blades tends to move them into low pitch (high rpm) in the absence of governor oil pressure. This is an exception to most of the aluminum hub models and feathering models. Most of the Hartzell propeller aluminum and steel hub models use centrifugal force acting on blade counterweights to increase blade pitch and governor oil pressure for low pitch. Many types of light aircraft use governor-regulated, constant-speed propellers in two-bladed and up to six-bladed versions. These propellers may be the nonfeathering type, or they may be capable of feathering and reversing. The steel hub contains a central "spider," that supports aluminum blades with a tube extending inside the blade roots. Blade clamps connect the blade shanks with blade retention bearings. A hydraulic cylinder is mounted on the rotational axis connected to the blade clamps for pitch actuation.

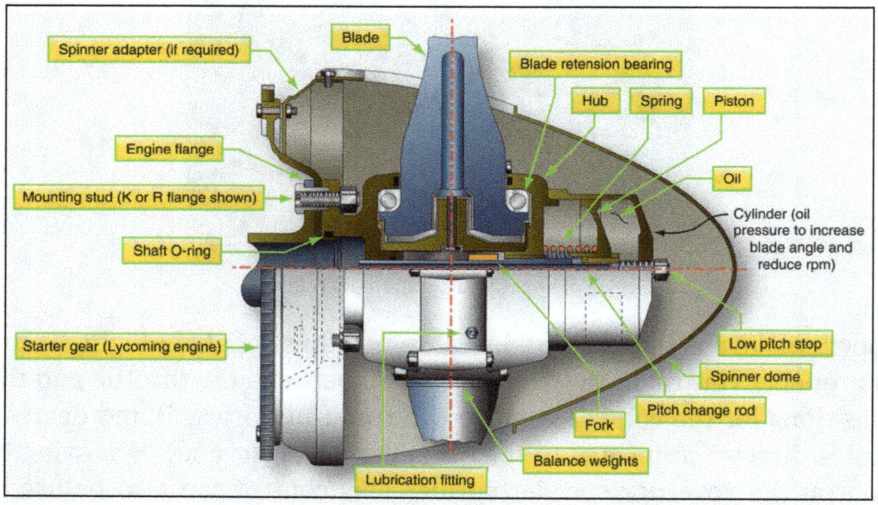

Constant speed non-feathering propeller.

The basic hub and blade retention is common to all models described. The blades are mounted on the hub spider for angular adjustment. The centrifugal force of the blades, amounting to as much as 25 tons, is transmitted to the hub spider through blade clamps and then through ball bearings. The propeller thrust and engine torque is transmitted from the blades to the hub spider through a bushing inside the blade shank. In order to control the pitch of the blades, a hydraulic piston-cylinder element is mounted on the front of the hub spider. The piston is attached to the blade clamps by means of a sliding rod and fork system for nonfeathering models and a link system for the feathering models. The piston is actuated in the forward direction by means of oil pressure supplied by a governor, which overcomes the opposing force created by the counterweights. Hartzell and McCauley propellers for light aircraft are similar in operation. The manufacturer's specifications and instructions must be consulted for information on specific models.

Constant-speed Feathering Propeller

The feathering propeller utilizes a single oil supply from a governing device to hydraulically actuate a change in blade angle. This propeller has five blades and is used primarily on Pratt & Whitney turbine engines. A two piece aluminum hub retains each propeller blade on a thrust bearing. A cylinder is attached to the hub and contains a feathering spring and piston. The hydraulically actuated piston transmits linear motion through a pitch change rod and fork to each blade to result in blade angle change.

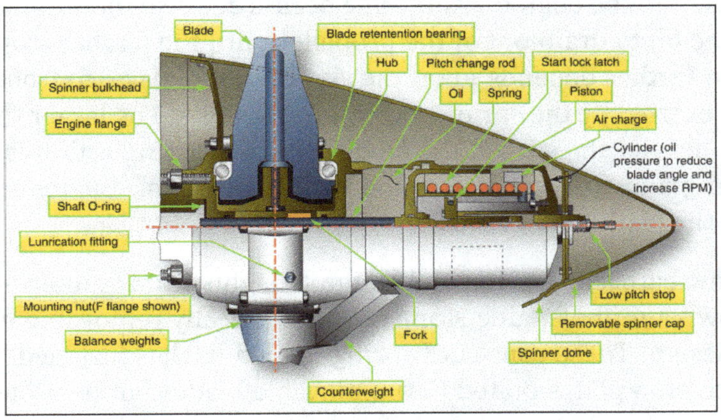

Constant-speed feathering propeller.

While the propeller is operating, the following forces are constantly present: 1) spring force, 2) counterweight force, 3) centrifugal twisting moment of each blade, and 4) blade aerodynamic twisting forces. The spring and counterweight forces attempt to rotate the blades to higher blade angle, while the centrifugal twisting moment of each blade is generally toward lower blade angle. Blade aerodynamic twisting force is usually very small in relation to the other forces and can attempt to increase or decrease blade angle. The summation of the propeller forces is toward higher pitch (low rpm) and is opposed by a variable force toward lower pitch (high rpm).

The variable force is oil under pressure from a governor with an internal pump that is mounted on and driven by the engine. The oil from the governor is supplied to the propeller and hydraulic piston through a hollow engine shaft. Increasing the volume of oil within the piston and cylinder decreases the blade angle and increases propeller rpm. If governor-supplied oil is lost during operation, the propeller increases pitch and feather. Feathering occurs because the summation of internal propeller forces causes the oil to drain out of the propeller until the feather stop position is reached. Normal in-flight feathering is accomplished when the pilot retards the propeller condition lever past the feather detent.

This permits control oil to drain from the propeller and return to the engine sump. Engine shutdown is normally accomplished during the feathering process. Normal in-flight unfeathering is accomplished when the pilot positions the propeller condition lever into the normal flight (governing) range and restarts the engine. As engine speed increases, the governor supplies oil to the propeller and the blade angle decreases. Decreasing the volume of oil increases blade angle and decrease propeller rpm. By changing blade angle, the governor can vary the load on the engine and maintain constant engine rpm (within limits), independent of where the power lever is set. The governor uses engine speed sensing mechanisms that permit it to supply or drain oil as necessary to maintain constant engine speed (rpm). Most of the steel hub Hartzell propellers and many of the aluminum hub are full feathering. These feathering propellers operate similarly to the nonfeathering ones except the feathering spring assists the counterweights to increase the pitch.

Feathering is accomplished by releasing the governor oil pressure, allowing the counterweights and feathering spring to feather the blades. This is done by pulling the condition lever (pitch control) back to the limit of its travel, which opens up a port in the governor allowing the oil from the propeller to drain back into the engine. Feathering occurs because the summation of internal propeller forces causes the oil to drain out of the propeller until the feather stop position is reached. The time necessary to feather depends upon the size of the oil passage from the propeller to the engine, and the force exerted by the spring and counterweights. The larger the passage is through the governor and the heavier the spring, the quicker the feathering action is. An elapsed time for feathering of between 3 and 10 seconds is usual with this system. Engine shutdown is normally accomplished during the feathering process.

In order to prevent the feathering spring and counterweights from feathering the propeller when the engine is shut down and the engine stopped, automatically removable high-pitch stops were incorporated in the design. These consist of spring-loaded latches fastened to the stationary hub that engage high-pitch stop plates bolted to the movable blade clamps. When the propeller is in rotation at speeds over 600–800 rpm, centrifugal force acts to disengage the latches from the high-pitch stop plates so that the propeller pitch may be increased to the feathering position. At lower rpm, or when the engine is stopped, the latch springs engage the latches with the high-pitch stops, preventing the pitch from increasing further due to the action of the feathering spring. The engine load would be excessive, especially on fixed-turbine turboprop engines. One safety feature inherent in this method of feathering is that the propeller feathers if the governor oil pressure drops to zero for any reason. As the governor obtains its supply of oil from the engine lubricating system, it follows that if the engine runs out of oil or if oil pressure fails due to breakage of a part of the engine, the propeller feathers automatically. This action may save the engine from further damage in case the pilot is not aware of trouble.

Unfeathering

Unfeathering can be accomplished by any of several methods, as follows:

- Start the engine, so the governor can pump oil back into the propeller to reduce pitch. In most light twins, this procedure is considered adequate since the feathering of the propeller would happen infrequently. Vibration can occur when the engine starts and the propeller starts to come out of feather.
- Provide an accumulator connected to the governor with a valve to trap an air-oil charge when the propeller is feathered, but released to the propeller when the rpm control is returned to normal position. This system is used with training aircraft because it unfeathers the propeller in a very short time and starts the engine wind milling.
- Provide an unfeathering pump that provides pressure to force the propeller back to low pitch quickly using engine oil.

Normal in-flight unfeathering is accomplished when the pilot positions the propeller condition lever into the normal flight (governing) range. This causes the governor to disconnect the propeller oil supply from drain and reconnects it to the governed oil supply line from the governor. At that point, there is no oil available from the engine oil pump to the governor; therefore, no governed oil is available from the governor for controlling the propeller blade angle and rpm. As the engine is started, its speed increases, the governor supplies oil to the propeller and the blade angle decreases. As soon as the engine is operating, the governor starts to unfeather the blades. Soon, windmilling takes place, which speeds up the process of unfeathering.

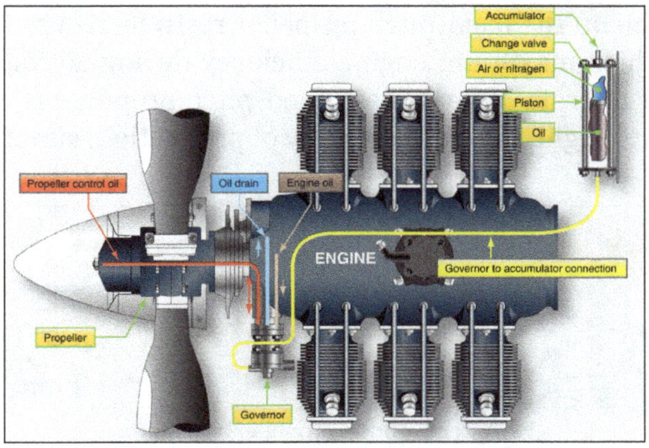

Unfeathering system.

In general, restarting and unfeathering of propellers can be classified as reciprocating engine restart unfeathering, turboprop engine restart unfeathering, and accumulator unfeathering. When reciprocating unfeathering is used, the engine takes a little longer to start turning enough to provide oil pressure to the governor and then to the propeller. This delay can cause vibration as the propeller is unfeathered. Many aircraft can use an accumulator to provide stored pressure to unfeather the propeller much quicker.

Special unfeathering systems are available for certain aircraft where restarting the engine is difficult or for training purposes. The system consists of an oil accumulator connected to the governor through a valve. The air or nitrogen pressure in one side of the accumulator pushes a piston to force oil from the other side of the accumulator through the governor to the propeller piston to move the propeller blades from feather to a lower blade angle. The propeller then begins to windmill and permits the engine to start. When the unfeathering pump is used, it is an additional pump that, once the propeller control is in the correct position, the (full increase rpm) pump is actuated and the oil pressure from the pump unfeathers the propeller.

Propellers used on General Aviation Aircraft

An increasing number of light aircraft are designed for operation with governor-regulated, constant-speed propellers. Significant segments of general aviation aircraft are still operated with fixed-pitch propellers. Light sport aircraft (LSA) use multi-blade fixed-pitch composite propellers on up to medium size turbo prop aircraft with reversing propeller systems. Larger transport and cargo turbo prop aircraft use propeller systems with dual or double-acting governors and differential oil pressure to change pitch.

Fixed-pitch Wooden Propellers

Although many of the wood propellers were used on older airplanes, some are still in use. The construction of a fixed pitch, wooden propeller is such that its blade pitch cannot be changed after manufacture. The choice of the blade angle is decided by the normal use of the propeller on an aircraft during level flight when the engine performs at maximum efficiency. The impossibility of changing the blade pitch on the fixed-pitch propeller restricts its use to small aircraft with low horsepower engines in which maximum engine efficiency during all flight conditions is of lesser importance than in larger aircraft. The wooden, fixed-pitch propeller is well suited for such small aircraft because of its light weight, rigidity, economy of production, simplicity of construction, and ease of replacement.

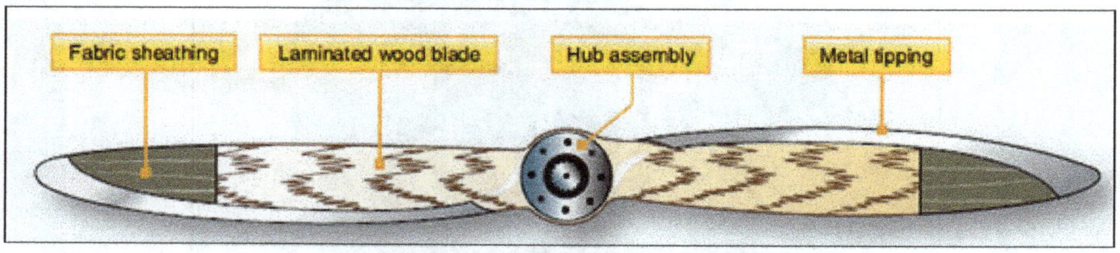

Fix-pitch wooden propeller assembly.

A wooden propeller is not constructed from a solid block, but is built up of a number of separate layers of carefully selected and well-seasoned hardwoods. Many woods, such as mahogany, cherry, black walnuts, and oak, are used to some extent, but birch is the most widely used. Five to nine separate layers are used, each about 3⁄4 inch thick. The several layers are glued together with waterproof, resinous glue and allowed to set. The blank is then roughed to the approximate shape and size of the finished product. The roughed-out propeller is then allowed to dry for approximately one week to permit the moisture content of the layers to become equalized. This additional period of seasoning prevents warping and cracking that might occur if the blank were immediately carved. Following this period, the propeller is carefully constructed. Templates and bench protractors are used to assure the proper contour and blade angle at all stations.

After the propeller blades are finished, a fabric covering is cemented to the outer 12 or 15 inches of each finished blade. A metal tipping is fastened to most of the leading edge and tip of each blade to protect the propeller from damage caused by flying particles in the air during landing, taxiing, or takeoff. Metal tipping may be of terneplate, Monel metal, or brass. Stainless steel has been used to some extent. It is secured to the leading edge of the blade by countersunk wood screws and rivets. The heads of the screws are soldered to the tipping to prevent loosening, and the solder is filed to make a smooth surface. Since moisture condenses on the tipping between the metal and the wood, the tipping is provided with small holes near the blade tip to allow this moisture to drain away or be thrown out by centrifugal force. It is important that these drain holes be kept open at all times. Since wood is subject to swelling, shrinking, and warping because of changes of moisture content, a protective coating is applied to the finished propeller to prevent a rapid change of moisture content. The finish most commonly used is a number of coats of water-repellent, clear varnish. After these processes are completed, the propeller is mounted on a spindle and very carefully balanced.

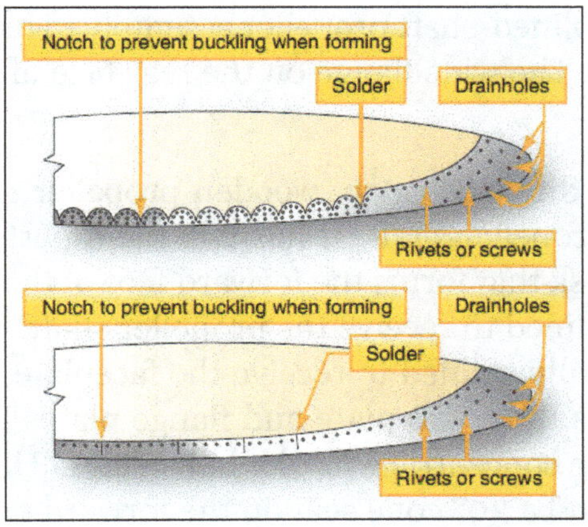

Installation of metal sheath and tipping.

Several types of hubs are used to mount wooden propellers on the engine crankshaft. The propeller may have a forged steel hub that fits a splined crankshaft; it may be connected to a tapered crankshaft by a tapered, forged steel hub; or it may be bolted to a steel flange forged on the crankshaft. In any case, several attaching parts are required to mount the propeller on the shaft properly.

Hubs fitting a tapered shaft are usually held in place by a retaining nut that screws onto the end of the shaft. On one model, a locknut is used to safety the retaining nut and to provide a puller for removing the propeller from the shaft. This nut screws into the hub and against the retaining nut. The locknut and the retaining nut are safe tied together with lock-wire or a cotter pin.

Front and rear cones may be used to seat the propeller properly on a splined shaft. The rear cone is a one-piece bronze cone that fits around the shaft and against the thrust nut (or spacer) and seats in the rear-cone seat of the hub.

The front cone is a two-piece, split-type steel cone that has a groove around its inner circumference so that it can be fitted over a flange of the propeller retaining nut. Then, the retaining nut is threaded into place and the front cone seats in the front cone hub. A snap ring is fitted into a groove in the hub in front of the front cone so that when the retaining nut is unscrewed from the propeller shaft, the front cone acts against the snap ring and pulls the propeller from the shaft.

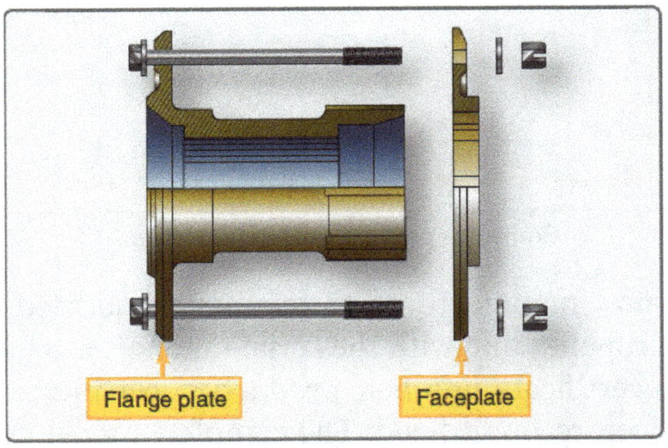

Hub assembly.

One type of hub incorporates a bronze bushing instead of a front cone. When this type of hub is used, it may be necessary to use a puller to start the propeller from the shaft. A rear cone spacer is sometimes provided with the splined-shaft propeller assembly to prevent the propeller from interfering with the engine cowling. The wide flange on the rear face of some types of hubs eliminates the use of a rear-cone spacer.

One type of hub assembly for the fixed-pitch, wooden propeller is a steel fitting inserted in the propeller to mount it on the propeller shaft. It has two main parts: the faceplate and the flange plate. The faceplate is a steel disk that forms the forward face of the hub. The flange plate is a steel flange with an internal bore splined to receive the propeller shaft. The end of the flange plate opposite the flange disk is externally splined to receive the faceplate; the faceplate bore has splines to match these external splines. Both faceplate and flange plates have a corresponding series of holes drilled on the disk surface concentric with the hub center. The bore of the flange plate has a 15° cone seat on the rear end and a 30° cone seat on the forward end to center the hub accurately on the propeller shaft.

Metal Fixed-Pitch

Propellers Metal fixed-pitch propellers are similar in general appearance to a wooden propeller, except that the sections are usually thinner. The metal fixed-pitch propeller is widely used on many models of light aircraft and LSA. Many of the earliest metal propellers were manufactured in one piece of forged Duralumin. Compared to wooden propellers, they were lighter in weight because of elimination of blade-clamping devices, offered a lower maintenance cost because they were made in one piece, provided more efficient cooling because of the effective pitch nearer the hub, and, because there was no joint between the blades and the hub, the propeller pitch could be changed, within limits, by twisting the blade slightly by a propeller repair station.

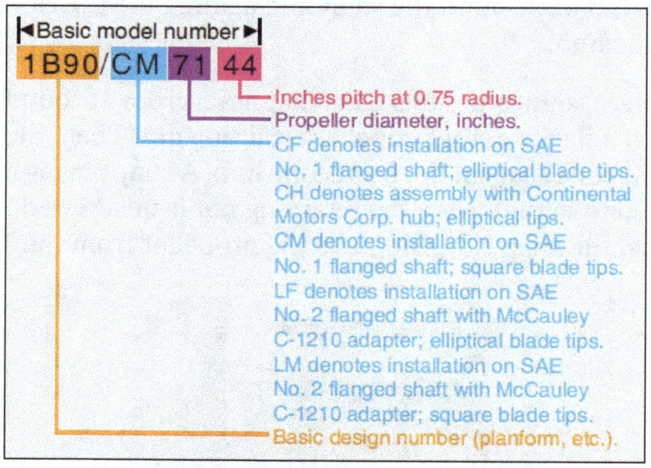

Complete propeller model numbers.

Propellers of this type are now manufactured as one-piece anodized aluminum alloy. They are identified by stamping the propeller hub with the serial number, model number, Federal Aviation Administration (FAA) type certificate number, production certificate number, and the number of times the propeller has been reconditioned. The complete model number of the propeller is a combination of the basic model number and suffix numbers to indicate the propeller diameter and pitch. An explanation of a complete model number, using the McCauley 1B90/CM propeller, is provided in figure.

Aircraft Propeller Vibration

Although vibration can be caused by the propeller, there are numerous other possible sources of vibration that can make troubleshooting difficult. If a propeller vibrates, whether due to balance, angle, or track problems, it typically vibrates throughout the entire rpm range, although the intensity of the vibration may vary with the rpm. If a vibration occurs only at one particular rpm or within a limited rpm range (e.g., 2200– 2350 rpm), the vibration is not normally a propeller problem but a problem of a poor engine-propeller match. If a propeller vibration is suspected but cannot be positively determined, the ideal troubleshooting method is to temporarily replace the propeller with one known to be airworthy and then test fly the aircraft if possible. Blade shake is not the source of vibration problems. Once the engine is running, centrifugal force holds the blades firmly (approximately 30,000–40,000 pounds) against blade bearings. Cabin vibration can sometimes be improved by reindexing the propeller to the crankshaft. The propeller can be removed, rotated 180°, and reinstalled. The propeller spinner can be a contributing factor to an out-of balance condition. An indication of this would be a noticeable spinner wobble while the engine is running. This condition is usually caused by inadequate shimming of the spinner front support or a cracked or deformed spinner.

When powerplant vibration is encountered, it is sometimes difficult to determine whether it is the result of engine vibration or propeller vibration. In most cases, the cause of the vibration can be determined by observing the propeller hub, dome, or spinner while the engine is running within a 1,200- to 1,500-rpm range, and determining whether or not the propeller hub rotates on an absolutely horizontal plane. If the propeller hub appears to swing in a slight orbit, the vibration is usually caused by the propeller. If the propeller hub does not appear to rotate in an orbit, the difficulty is probably caused by engine vibration.

When propeller vibration is the reason for excessive vibration, the difficulty is usually caused by propeller blade imbalance, propeller blades not tracking, or variation in propeller blade angle settings. Check the propeller blade tracking and then the low-pitch blade angle setting to determine if either is the cause of the vibration. If both propeller tracking and low blade angle setting are correct, the propeller is statically or dynamically unbalanced and should be replaced, or re-balanced if permitted by the manufacturer.

Blade Tracking

Blade tracking is the process of determining the positions of the tips of the propeller blades relative to each other (blades rotating in the same plane of rotation). Tracking shows only the relative position of the blades, not their actual path. The blades should all track one another as closely as possible. The difference in track at like points must not exceed the tolerance specified by the propeller manufacturer. The design and manufacture of propellers is such that the tips of the blades give a good indication of tracking. The following method for checking tracking is normally used:

- Chock the aircraft so it cannot be moved.
- Remove one spark plug from each cylinder. This makes the propeller easier and safer to turn.

Rotate one of the blades so it is pointing down.

- Place a solid object (e.g., a heavy wooden block that is at least a couple of inches higher off the ground than the distance between the propeller tip and the ground) next to the propeller tip so that it just touches or attaches a pointer/indicator to the cowling itself.
- Rotate the propeller slowly to determine if the next blade tracks through the same point (touches the block/pointer). Each blade track should be within 1/16 inch (plus or minus) from the opposite blade's track.
- An out-of-track propeller may be due to one or more propeller blades being bent, a bent propeller flange, or propeller mounting bolts that are either over- or under torqued. An out-of-track propeller causes vibration and stress to the airframe and engine and may cause premature propeller failure.

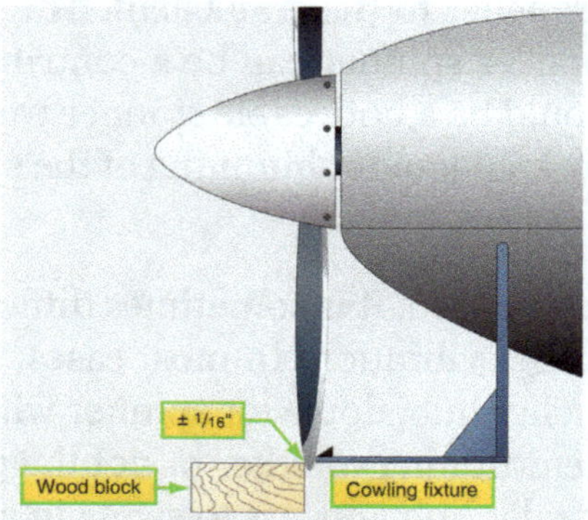

Propeller blade tracking.

Checking and Adjusting Propeller Blade Angles

When you find an improper blade angle setting during installation or is indicated by engine performance, follow basic maintenance guidelines. From the applicable manufacturer's instructions, obtain the blade angle setting and the station at which the blade angle is checked. Do not use metal scribes or other sharply pointed instruments to mark the location of blade stations or to make reference lines on propeller blades, since such surface scratches can eventually result in blade failure. Use a bench-top protractor if the propeller is removed from the aircraft. Use a handheld protractor (a digital protractor provides an easy measurement) to check blade angle if the propeller is installed on the aircraft or is placed on the knife-edge balancing stand.

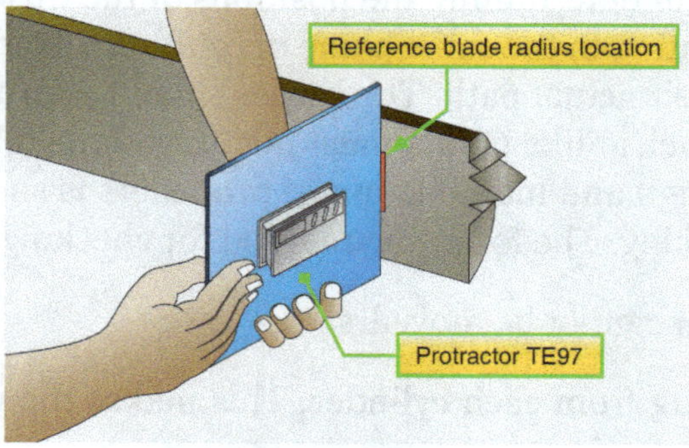

Blade angle measurement.

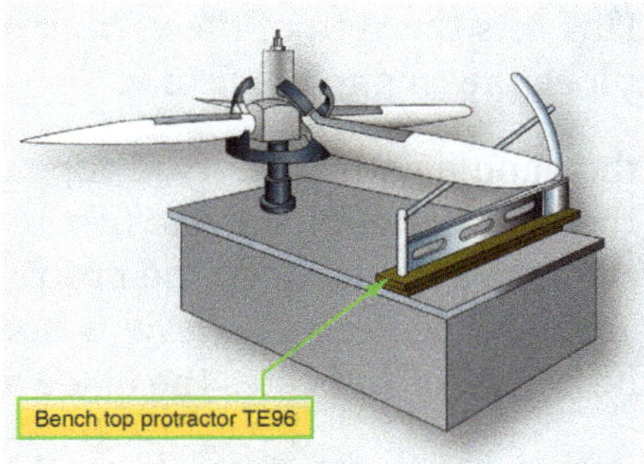

Bench top protractor.

Universal Propeller Protractor

The universal propeller protractor can be used to check propeller blade angles when the propeller is on a balancing stand or installed on the aircraft engine. Figure shows the parts and adjustments of a universal propeller protractor. The following instructions for using the protractor apply to a propeller installed on the engine.

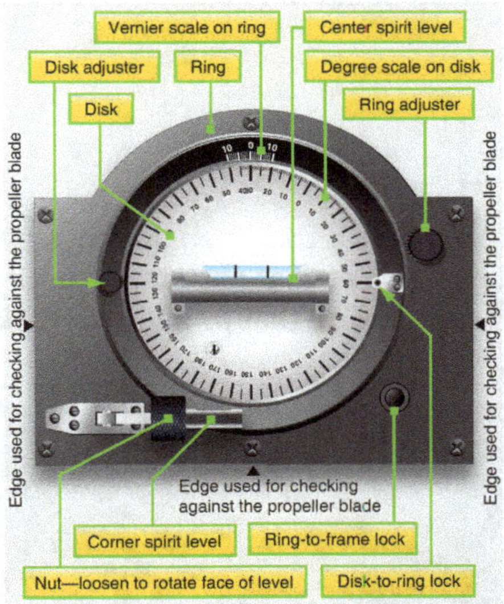

Universal proptractor.

Turn the propeller until the first blade to be checked is horizontal with the leading edge up. Place the corner spirit level at right angles to the face of the protractor. Align degree and vernier scales by turning the disk adjuster before the disk is locked to the ring. The locking device is a pin that is held in the engaged position by a spring. The pin can be released by pulling it outward and turning it 90°.

Release the ring-to-frame lock (a right-hand screw with thumb nut) and turn the ring until both ring and disk zeros are at the top of the protractor.

Check the blade angle by determining how much the flat side of the block slants from the plane of rotation. First, locate a point to represent the plane of rotation by placing the protractor vertically against the end of the hub nut or any convenient surface known to lie in the plane of propeller rotation. Keep the protractor vertical by the corner spirit level, and turn the ring adjuster until the center spirit level is horizontal. This sets the zero of the vernier scale at a point representing the plane of propeller rotation. Then, lock the ring to the frame.

While holding the protractor by the handle with the curved edge up, release the disk-to-ring lock. Place the forward vertical edge (the edge opposite the one first used) against the blade at the station specified in the manufacturer's instructions. Keep the protractor vertical by the corner spirit level, and turn the disk adjuster until the center spirit level is horizontal. The number of degrees and tenths of a degree between the two zeros indicates the blade angle.

In determining the blade angle, remember that ten points on the vernier scale are equal to nine points on the degree scale. The graduations on the vernier scale represent tenths of a degree, but those of the degree scale represent whole degrees. The number of tenths of a degree in the blade angle is given by the number of vernier scale spaces between the zero of the vernier scale and the vernier scale graduation line nearest to perfect alignment with a degree scale graduation line. This reading should always be made on the vernier scale. The vernier scale increases in the same direction that the protractor scale increases. This is opposite to the direction of rotation of the moving element of the protractor. After making any necessary adjustment of the blade, lock it in position and repeat the same operations for the remaining blades of the propeller.

Rocket Propellant

Rocket propellant is the material used by a rocket engine to generate thrust. The most common propellants consist of a fuel and an oxidizer that participate in chemical reactions to produce extremely hot gases. These gases exert pressure that propels the rocket forward while they are ejected in the form of a jet through the rear.

Space Shuttle Atlantis during takeoff.

The development of extremely powerful rocket propellants has enabled the launching of Earth-orbiting satellites, ballistic missiles, manned spacecraft, and interplanetary space probes.

Principles of Operation

In a chemically powered rocket, the engine creates thrust (forward force) by combustion of the propellant materials to form very hot gases, which expand in the combustion chamber and are ejected as a high-speed jet through a nozzle in the rear.

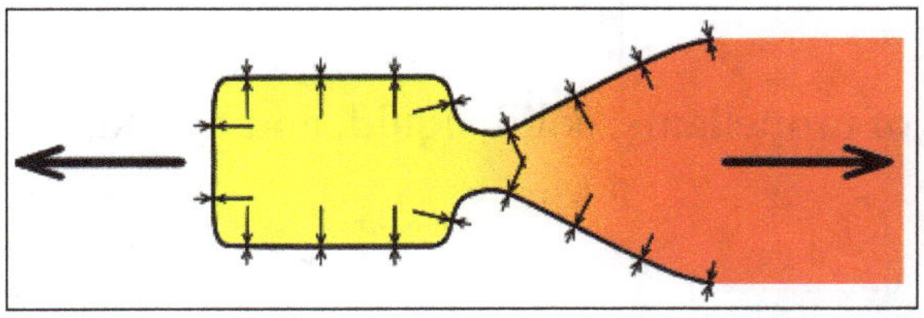

Rocket thrust is produced by pressures acting on the combustion chamber and nozzle.

In a closed chamber, the gas pressure would be equal in each direction and no acceleration would occur. By providing an opening at the bottom of the chamber, there is no pressure acting on that side, but exhaust escapes from that end. The remaining components of pressure produce a thrust on the side opposite the opening. Using a nozzle increases the forces further, in fact multiplies the thrust as a function of the area ratio of the nozzle, because the pressures also act on the nozzle. In addition, the pressures act on the exhaust in the opposite direction and accelerate it to very high speeds (in accordance with Newton's Third Law of motion). This disequilibrium of pressures can be maintained for as long as propellant is added to the combustion chamber.

It turns out (from the conservation of momentum) that the speed of the exhaust of a rocket determines how much momentum increase is created for a given amount of propellant. The faster the net speed of the exhaust in one direction, the greater the speed of the rocket in the opposite direction can become. As the propellant supply decreases, the vehicle becomes lighter and acceleration increases until the rocket eventually runs out of propellant. Consequently, much of the speed change occurs toward the end of the burn when the vehicle is much lighter.

These operational principles stand in contrast to the commonly held assumption that a rocket "pushes" against the air beneath it. Rockets perform better in space, where there is practically nothing behind them to push against, because they do not need to overcome air resistance and atmospheric pressure.

The maximum velocity that a rocket can attain in the absence of any external forces is primarily a function of its mass ratio and its *exhaust velocity*. The relationship is described by the following rocket equation:

$$V_f = V_e \ln(M_0 / M_f)$$

The mass ratio is a way to express what proportion of the rocket is fuel when it starts accelerating. Typically, a single-stage rocket might have a mass fraction of 90 percent propellant, which is a mass ratio of 1/(1-0.9) = 10. The exhaust velocity is often reported as the rocket's *specific impulse*.

The first stage of a rocket usually uses high-density (low-volume) propellants to reduce the area exposed to atmospheric drag and to obtain lighter tankage and higher thrust/weight ratios. Thus, the Apollo Saturn V first stage used kerosene-liquid oxygen rather than the liquid hydrogen-liquid oxygen used on the upper stages. (Hydrogen is highly energetic per kilogram, but not per cubic meter). Similarly, the Space Shuttle uses high-thrust, high-density solid rocket boosters (SRBs) for its lift-off, with liquid hydrogen-liquid oxygen used partly for lift-off but primarily for orbital insertion of the shuttle.

Chemical Propellants

There are three main types of propellants: solid, liquid, and hybrid.

Solid Propellants

The earliest rockets were created hundreds of years ago by the Chinese, and were used primarily for fireworks displays and as weapons. They were fueled with black powder, a type of gunpowder consisting of a mixture of charcoal, sulfur and potassium nitrate (saltpeter). Rocket propellant technology did not advance until the end of the nineteenth century, by which time smokeless powder had been developed, originally for use in firearms and artillery pieces. Smokeless powders and related compounds have seen use as double-base propellants.

Solid propellants (and almost all rocket propellants) consist of an oxidizer and a fuel. In the case of gunpowder, the fuel is charcoal, the oxidizer is potassium nitrate, and sulfur serves as a catalyst. (Sulfur is not a true catalyst in gunpowder as it is consumed to a great extent into a variety of reaction products such as K_2S. The sulfur acts mainly as a sensitizer lowering threshold of ignition).

However, solid-fuel rockets have a number of disadvantages relative to liquid-fuel rockets. Solid rockets have a lower specific impulse than liquid fueled rockets. It is also difficult to build a large mass ratio solid rocket because almost the entire rocket is the combustion chamber, and must be built to withstand the high combustion pressures. If a solid rocket is used to go all the way to orbit, the payload fraction is very small. (For example, the Orbital Sciences Pegasus rocket is an air-launched three-stage solid rocket orbital booster. Launch mass is 23,130 kg, low earth orbit payload is 443 kg, for a payload fraction of 1.9 percent. Compare that to a Delta IV Medium, 249,500 kg, payload 8600 kg, payload fraction 3.4 percent without air-launch assistance).

A drawback to solid-fuel rockets is that they cannot be throttled in real time, although a pre-designed thrust schedule can be built into the grain during manufacture.

Solid-fuel rockets can often be shut down before they run out of fuel. Essentially, the rocket is vented or an extinguishing injected so as to terminate the combustion process. In some cases, termination destroys the rocket, and this is typically done only by a Range Safety Officer if the rocket goes awry. The third stages of the Minuteman and MX rockets have precision shutdown ports which, when opened, reduce the chamber pressure so abruptly that the interior flame is blown out. This allows a more precise trajectory which improves targeting accuracy.

Finally, casting very large single-grain rocket motors has proved to be a very tricky business. Defects in the grain can cause explosions during the burn, and these explosions can increase the burning propellant surface enough to cause a runaway pressure increase, until the case fails.

Liquid Propellants

Liquid-fueled rockets have better specific impulse than solid rockets and are capable of being throttled, shut down, and restarted. Only the combustion chamber of a liquid-fueled rocket needs to withstand combustion pressures and temperatures. On vehicles employing turbo pumps, the fuel tanks carry very much less pressure and thus can be built far more lightly, permitting a larger mass ratio. For these reasons, most orbital launch vehicles and all first- and second-generation ICBMs use liquid fuels for most of their velocity gain.

The primary performance advantage of liquid propellants is the oxidizer. Several practical liquid oxidizers (liquid oxygen, nitrogen tetroxide, and hydrogen peroxide) are available which have much better specific impulse than ammonium perchlorate when paired with comparable fuels.

Most liquid propellants are also cheaper than solid propellants. For orbital launchers, the cost savings do not and historically have not mattered; the cost of fuel is a very small portion of the overall cost of the rocket, even in the case of solid fuels.

The main difficulties with liquid propellants are also with the oxidizers. The oxidizers are generally at least moderately difficult to store and handle due to their high reactivity with common materials, and they may have extreme toxicity (nitric acids) or moderately cryogenic properties (liquid oxygen or "LOX") or both (liquid fluorine, FLOX-a fluorine/LOX mix). Several exotic oxidizers have been proposed: Liquid ozone (O_3), ClF_3, and ClF_5, all of which are unstable, energetic, and toxic.

Liquid-fueled rockets also require potentially troublesome valves and seals and thermally stressed combustion chambers, which increase the cost of the rocket. Many employ specially designed turbo pumps which raise the cost enormously due to difficult fluid flow patterns that exist within the casings.

Though all the early rocket theorists proposed liquid hydrogen and liquid oxygen as propellants, the first liquid-fueled rocket, launched by Robert Goddard on March 16, 1926, used gasoline and liquid oxygen. Liquid hydrogen was first used by the engines designed by Pratt and Whitney for the Lockheed CL-400 Suntan reconnaissance aircraft in the mid-1950s. In the mid-1960s, the Centaur and Saturn upper stages were both using liquid hydrogen and liquid oxygen.

The highest specific impulse chemistry ever test-fired in a rocket engine was lithium and fluorine, with hydrogen added to improve the exhaust thermodynamics (making this a tripropellant). The combination delivered 542 seconds (5.32 kN·s/kg, 5320 m/s) specific impulse in a vacuum. The impracticality of this chemistry highlights why exotic propellants are not actually used: to make all three components liquids, the hydrogen must be kept below -252 °C (just 21 K) and the lithium must be kept above 180 °C (453 K). Lithium and fluorine are both extremely corrosive, lithium ignites on contact with air, fluorine ignites on contact with most fuels, and hydrogen, while not hypergolic, is an explosive hazard. Fluorine and the hydrogen fluoride (HF) in the exhaust are very toxic, which damages the environment, makes work around the launch pad difficult, and makes getting a launch license that much more difficult. The rocket exhaust is also ionized, which would interfere with radio communication with the rocket.

Common liquid propellant combinations currently in use are:

- LOX and kerosene (RP-1) used for the lower stages of most Russian and Chinese boosters, the first stages of the Saturn V and Atlas V, and all stages of the developmental Falcon 1 and Falcon 9. Very similar to Robert Goddard's first rocket. This combination is widely regarded as the most practical for civilian orbital launchers.

- LOX and liquid hydrogen used in the Space Shuttle, the Centaur upper stage, the newer Delta IV rocket, the H-IIA rocket, and most stages of the European Ariane rockets.
- Nitrogen tetroxide (N_2O_4) and hydrazine (N_2H_4), MMH, or UDMH. Used in military, orbital and deep space rockets, because both liquids are storable for long periods at reasonable temperatures and pressures. This combination is hypergolic, making for attractively simple ignition sequences. The major inconvenience is that these propellants are highly toxic, hence they require careful handling. Hydrazine also decomposes energetically to nitrogen, hydrogen, and ammonia, making it a fairly good monopropellant.

Gas Propellants

A gas propellant usually involves some sort of compressed gas. However, due to the low density and high weight of the pressure vessel, gases are seldom used at present.

Hybrid Propellants

A hybrid rocket usually has a solid fuel and a liquid or gas oxidizer. The fluid oxidizer can make it possible to throttle and restart the motor just like a liquid-fueled rocket. Hybrid rockets are also cleaner than solid rockets because practical high-performance solid-phase oxidizers all contain chlorine, versus the more benign liquid oxygen or nitrous oxide used in hybrids. Because just one propellant is a fluid, hybrids are simpler than liquid rockets.

Hybrid motors suffer two major drawbacks. The first, shared with solid rocket motors, is that the casing around the fuel grain must be built to withstand full combustion pressure and often extreme temperatures as well. However, modern composite structures handle this problem well, and when used with nitrous oxide or hydrogen peroxide relatively small percentage of fuel is needed anyway, so the combustion chamber is not especially large.

The primary remaining difficulty with hybrids is with mixing the propellants during the combustion process. In solid propellants, the oxidizer and fuel are mixed in a factory in carefully controlled conditions. Liquid propellants are generally mixed by the injector at the top of the combustion chamber, which directs many small swift-moving streams of fuel and oxidizer into one another. Liquid fueled rocket injector design has been studied at great length and still resists reliable performance prediction. In a hybrid motor, the mixing happens at the melting or evaporating surface of the fuel. The mixing is not a well-controlled process and generally quite a lot of propellant is left unburned, which limits the efficiency and thus the exhaust velocity of the motor. Additionally, as the burn continues, the hole down the center of the grain (the "port") widens and the mixture ratio tends to become more oxidizer rich.

Inert Propellants

Some rocket designs have their propellants obtain their energy from non-chemical or even external sources. For example, water rockets use the compressed gas, typically air, to force the water out of the rocket.

Solar thermal rockets and nuclear thermal rockets typically propose to use liquid hydrogen for an I_{sp} (Specific Impulse) of around 600-900 seconds, or in some cases water that is exhausted as steam for an I_{sp} of about 190 seconds.

Additionally, for low performance requirements such as attitude jets, inert gases such as nitrogen have been employed.

Mixture Ratio

The theoretical exhaust velocity of a given propellant chemistry is a function of the energy released per unit of propellant mass (specific energy). Unburned fuel or oxidizer drags down the specific energy. Surprisingly, most rockets run fuel-rich.

The usual explanation for fuel-rich mixtures is that fuel-rich mixtures have lower molecular weight exhaust, which by reducing M supposedly increases the ratio $\frac{\sqrt{T_c}}{M}$, which is approximately equal to the theoretical exhaust velocity. This explanation, though found in some textbooks, is wrong. Fuel-rich mixtures actually have lower theoretical exhaust velocities, because $\sqrt{T_c}$ decreases as fast as or faster than M.

The nozzle of the rocket converts the thermal energy of the propellants into directed kinetic energy. This conversion happens in a short time, on the order of one millisecond. During the conversion, energy must transfer very quickly from the rotational and vibrational states of the exhaust molecules into translation. Molecules with fewer atoms (like CO and H_2) store less energy in vibration and rotation than molecules with more atoms (like CO_2 and H_2O). These smaller molecules transfer more of their rotational and vibrational energy to translation energy than larger molecules, and the resulting improvement in nozzle efficiency is large enough that real rocket engines improve their actual exhaust velocity by running rich mixtures with somewhat lower theoretical exhaust velocities.

The effect of exhaust molecular weight on nozzle efficiency is most important for nozzles operating near sea level. High expansion rockets operating in a vacuum see a much smaller effect, and so are run less rich. The Saturn-II stage (a LOX/LH_2 rocket) varied its mixture ratio during flight to optimize performance.

LOX/hydrocarbon rockets are run only somewhat rich (O/F mass ratio of 3 rather than stoichiometric of 3.4 to 4), because the energy release per unit mass drops off quickly as the mixture ratio deviates from stoichiometric. LOX/LH_2 rockets are run very rich (O/F mass ratio of 4 rather than stoichiometric 8) because hydrogen is so light that the energy release per unit mass of propellant drops very slowly with extra hydrogen. In fact, LOX/LH_2 rockets are generally limited in how rich they run by the performance penalty of the mass of the extra hydrogen tankage, rather than the mass of the hydrogen itself.

Another reason for running rich is that off-stoichiometric mixtures burn cooler than stoichiometric mixtures, which makes engine cooling easier. And as most engines are made of metal or carbon, hot oxidizer-rich exhaust is extremely corrosive, where fuel-rich exhaust is less so. American engines have all been fuel-rich. Some Soviet engines have been oxidizer-rich.

Additionally, there is a difference between mixture ratios for optimum I_{sp} and optimum thrust. During launch, shortly after takeoff, high thrust is at a premium. This can be achieved at some temporary reduction of I_{sp} by increasing the oxidizer ratio initially, and then transitioning to more fuel-rich mixtures. Since engine size is typically scaled for takeoff thrust this permits reduction of the weight of rocket engine, pipes and pumps and the extra propellant use can be more than compensated by increases of acceleration towards the end of the burn by having a reduced dry mass.

Propellant Density

Although liquid hydrogen gives a high I_{sp}, its low density is a significant disadvantage: hydrogen occupies about 7x more volume per kilogram than dense fuels such as kerosene. This not only penalizes the tankage, but also the pipes and fuel pumps leading from the tank, which need to be 7x bigger and heavier. (The oxidizer side of the engine and tankage is of course unaffected.) This makes the vehicle's dry mass much higher, so the use of liquid hydrogen is not such a big win as might be expected. Indeed, some dense hydrocarbon/LOX propellant combinations have higher performance when the dry mass penalties are included.

Due to lower I_{sp}, dense propellant launch vehicles have a higher takeoff mass, but this does not mean a proportionately high cost; on the contrary, the vehicle may well end up cheaper. Liquid hydrogen is quite an expensive fuel to produce and store, and causes many practical difficulties with design and manufacture of the vehicle.

Because of the higher overall weight, a dense-fueled launch vehicle necessarily requires higher takeoff thrust, but it carries this thrust capability all the way to orbit. This, in combination with the better thrust/weight ratios, means that dense-fueled vehicles reach orbit earlier, thereby minimizing losses due to gravity drag. Thus, the effective delta-v requirements for these vehicles are reduced.

However, liquid hydrogen does give clear advantages when the overall mass needs to be minimized; for example the Saturn V vehicle used it on the upper stages; this reduced weight meant that the dense-fueled first stage could be made proportionately smaller, saving quite a bit of money.

References

- Propeller-aerodynamic-process-airplane: aircraftsystemstech.com, Retrieved 17 March, 2019
- Constant-speed-propellers-hartzell: aircraftsystemstech.com, Retrieved 13 February, 2019
- Propellers-used-on-general-aviation-aircraft: flight-mechanic.com, Retrieved 7 May, 2019
- Propeller-vibration: aircraftsystemstech.com, Retrieved 13 July, 2019
- Rocket-propellant, entry: newworldencyclopedia.org, Retrieved 3 June, 2019

5

Aircraft Instruments and Systems

There are numerous instruments and systems which are used in aircraft such as pitot-static systems, compass systems, gyroscopic flight instruments, navigational systems and flight control systems. This chapter discusses in detail these systems and the different instruments which are a part of these systems.

Pitot-static Systems

An aircraft pitot-static system comprises a number of sensors which detect the ambient air pressure affected (pitot pressure) and unaffected (static pressure) by the forward motion of the aircraft. These pressures are used on their own or in combination with each other to provide indications of various flight parameters.

These include:

- Altitude (Altimeter).
- Airspeed (Air Speed Indicator).
- Mach Number (Machmeter).
- Vertical speed (Vertical Speed Indicator).

Pitot and static pressure are also used in other equipment, such as the Autopilot and the Cabin Altimeter.

Static Pressure

Static pressure is measured through a number of vents, situated at aerodynamically neutral points on the aircraft fuselage. Vents are sited on either side of the fuselage and feed into a common tube; this has the effect of cancelling out to some extent errors arising from the position of the vents.

A combination of careful vent siting and accurate calibration reduces errors to an acceptable degree. Commercial aircraft have at least two completely independent static systems to provide redundancy in the case of system failure.

Static vents are often plugged when the aircraft is parked for more than a short period of time to reduce the chance of blockage or contamination. Vents may be electrically heated to prevent blockage by ice.

Pitot Pressure

Pitot pressure is measured in a pitot tube or pressure head, which is an open tube facing forward along the axis of the aircraft. The pressure measured in the tube is a combination of static pressure and pressure due to the aircraft forward speed. Pitot tubes are carefully sited to reduce to a minimum error due to the flow of air over the aircraft.

Commercial aircraft have at least two completely independent pitot systems to provide redundancy in the case of system failure.

Pitot tubes are normally covered when the aircraft is parked for more than a short period of time to reduce the chance of blockage or contamination. They are invariably electrically heated to reduce contamination by moisture and prevent blockage by ice.

Airspeed Indicator

The Airspeed Indicator (ASI) is a Pitot-static instrument used in an aircraft to display the crafts airspeed, typically in knots. Airspeed indication is accomplished with the use of a thin, corrugated phosphor bronze diaphragm (aneroid) which measures Dynamic Pressure of the air between the Pitot tube (ram air) and static port (static pressure).

- Dynamic Pressure: Difference between the static (ambient) air pressure and the total pressure caused by the motion of the aircraft through the air.

The primary use of the airspeed indicator is to provide performance guidance during climb, descent, and landing.

Working Principle of Airspeed Indicator

Airspeed is a measure of differential pressure between the pitot (impact/dynamic pressure) and static pressures. Simply stated, ram air is pushed against a diaphragm, which is compared to the static pressure.

Pitot Tube.

Static Port.

The static pressure is captured through the static port(s) located on the side of the fuselage:

- The location is chosen at a location to most accurately detect prevailing atmospheric pressure (parallel to air stream) and avoid dynamic (ram) air pressure.
- Some aircraft will have more than one port to more accurately measure pressure during slips and skids.

"Ram air" is the air captured through the opening of the Pitot tube by the passage of the aircraft through the air:

- Ram Air can also be termed as total pressure.
- Some Pitot tubes are electrically heated to prevent clogging with ice.
- Most aircraft have an alternate static source intended for use when the primary static source is blocked and is especially important when in instrument meteorological conditions (IMC).
    - Alternate static sources are typically less accurate.

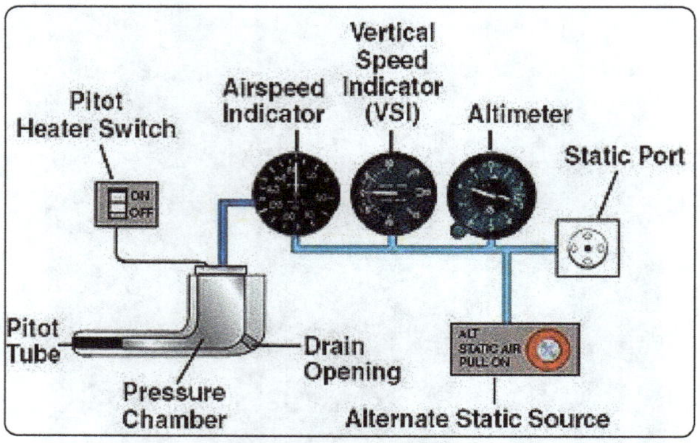

Pitot-Static System.

Conservation of Energy states that total pressure must remain the same and therefore as the Pitot pressure increases or the static pressure decreases, the diaphragm expands. This dimensional change is measured by a rocking shaft and a set of gears that drives a pointer across the instrument dial.

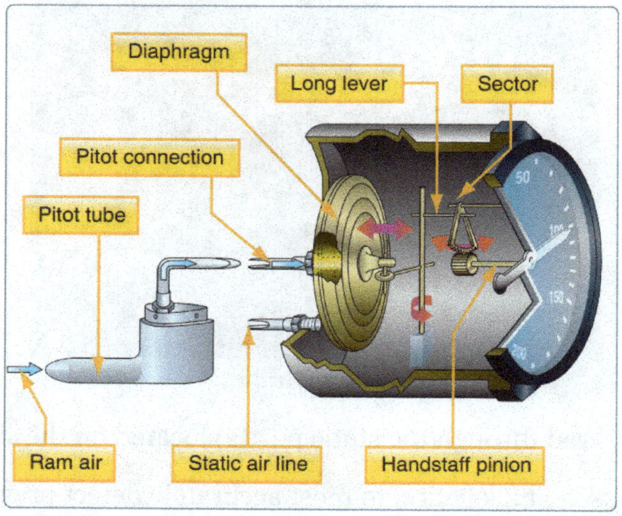

Pilot Handbook of Aeronautical Knowledge, Airspeed indicator (ASI).

Types of Airspeeds

Aircrew are primarily concerned with Indicated Airspeed and True Airspeed in flight with regards to performance.

Indicated Airspeed

The IAS is the direct airspeed reading shown by an airspeed indicator. The reading has not been corrected for variations in atmospheric density, installation error, or instrument errors. As height increases, the indicated airspeed falls below the true airspeed. Manufacturers use this airspeed as the basis for determining aircraft performance. IAS will not normally vary with altitude or temperature and so your V-speeds listed in the AFM/POH will mostly vary due to weight.

FLAPS UP											
KIAS	40	50	60	70	80	90	100	110	120	130	140
KCAS	46	53	60	69	78	88	97	107	117	127	136
FLAPS 10°											
KIAS	40	50	60	70	80	85	---	---	---	---	---
KCAS	44	52	61	70	80	84	---	---	---	---	---
FLAPS 30°											
KIAS	40	50	60	70	80	85	---	---	---	---	---
KCAS	43	51	61	71	82	87	---	---	---	---	---

Indicated to calibrated airspeed conversion chart example.

Calibrated Airspeed

CAS is the indicated airspeed of an aircraft, corrected for position and instrument error:

- Errors can include angle of attack, flap configuration, ground proximity, wind direction, to name a few.
- Errors can sometimes equal several knots and are generally greatest at low airspeeds.
- Any errors that interfere with the system reading total and static pressure (which when subtracted give you dynamic pressure) are corrected here.
- This will give the actual speed in which aircraft is moving through the air.
- Calibrated airspeed is equal to true airspeed in standard atmosphere at sea level (High AoA, minimal error at cruise).
- The POH/AFM has a chart or graph to correct IAS for these errors and provide the correct CAS for the various flap and landing gear configurations.

Some aircraft have alternate static sources which may need to be referenced in a separate chart.

Equivalent Airspeed

Equivalent airspeed is more designed for engineers and not usually practical for pilots. The airspeed corrected for compressibility effects above 180-200 knots and 20,000', which is the airspeed the airplane "feels". As the airspeed and pressure altitude increase, the CAS becomes higher than it should be as air molecules begin to stack up against the aircraft and instruments. A correction for compression must be subtracted from the CAS.

True Airspeed

Because the Pitot-system does not detect air density changes, it is calibrated to standard sea level pressure and any change in pressure (or altitude) thereby requires a correction.

Further, as air density decreases with an increase in altitude, an aircraft has to be flown faster at higher altitudes to cause the same pressure difference between pitot impact pressure and static pressure.

- Therefore, for a given CAS, TAS increases as altitude increases; or for a given TAS, CAS decreases as altitude increases.

TAS is therefore CAS corrected for non-standard temperature, with the help of an Outside Air Temperature (OAT) gauge, and altitude. The TAS is the speed that is used for flight planning and is used when filing a flight plan. On higher performance aircraft, a true airspeed indicator may be installed.

- Calculating True Airspeed:
  - Flight Computer Method: The most accurate method is to use a flight computer. With this method, the CAS is corrected for temperature and pressure variation by using the airspeed correction scale on the computer. Extremely accurate electronic flight computers are also available. Just enter the CAS, pressure altitude, and temperature, and the computer calculates the TAS.
  - Rule of Thumb: A second method, which is a rule of thumb, provides the approximate TAS. Simply add 2 percent to the CAS for each 1,000 feet of altitude. Formula:
    - 5 (5000 ft) * 0.02 = .1 (correction factor).
    - 1 * 100 KCAS (cruise airspeed) = 10 knots (correction speed).
    - 100 (CAS) + 10 = 110 knots TAS.

Ground Speed

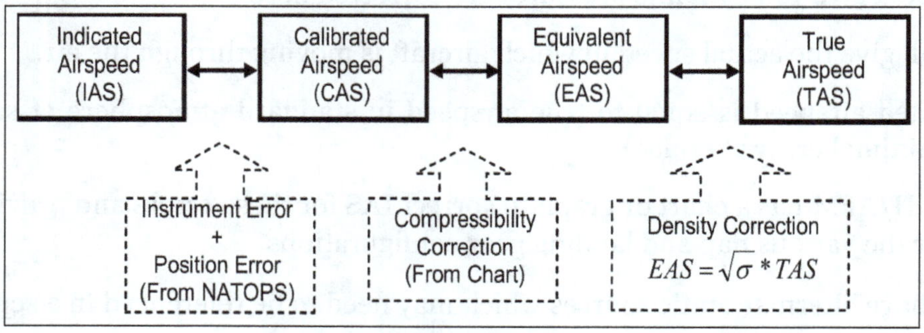

Airspeed Conversions.

Groundspeed (GS) is the actual speed of the airplane over the ground. It is TAS adjusted for wind (airmass movement). GS decreases with a headwind and increases with a tailwind. Ground Speed is a primary concern for performance during cross-country planning.

Instrument Flying Handbook, A maximum allowable airspeed indicator has a movable pointer that indicates the never-exceed speed, which changes with altitude to avoid the onset of transonic shock waves.

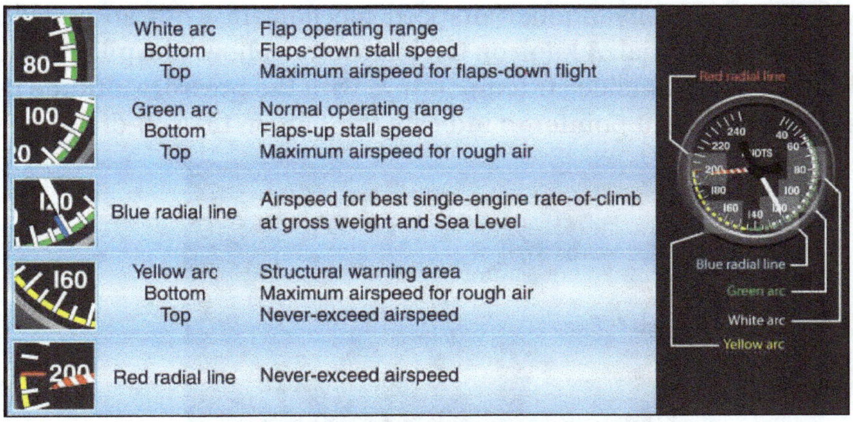

Airspeed Indicator Markings.

Mach Number

Mach number is the ratio of the TAS of the aircraft to the speed of sound in the same atmospheric conditions. Some older mechanical Machmeters not driven from an air data computer use an altitude aneroid inside the instrument that converts pitot-static pressure into Mach number. Modern electronic Machmeters use information from an air data computer system to correct for temperature errors to display true Mach number.

Maximum Allowable Airspeed

The maximum airspeed pointer is actuated by an aneroid, or altimeter mechanism, that moves it to a lower value as air density decreases. This instrument looks much like a standard air-speed indicator, calibrated in knots, but has an additional pointer colored red, checkered, or striped. The maximum airspeed pointer is actuated by an aneroid, or altimeter mechanism, that moves it to a lower value as air density decreases.

Instrument Flying Handbook, A Machmeter shows the ratio of the speed of sound to the TAS the aircraft is flying.

Some aircraft are equipped with true ASIs that have a temperature-compensated aneroid bellows inside the instrument case. This bellows modifies the movement of the rocking shaft inside the instrument case so the pointer shows the actual TAS.

These instruments have the conventional airspeed mechanism, with an added sub-dial visible through cutouts in the regular dial. A knob on the instrument allows the pilot to rotate the sub-dial and align an indication of the outside air temperature with the pressure altitude being flown. This alignment causes the instrument pointer to indicate the TAS on the sub-dial.

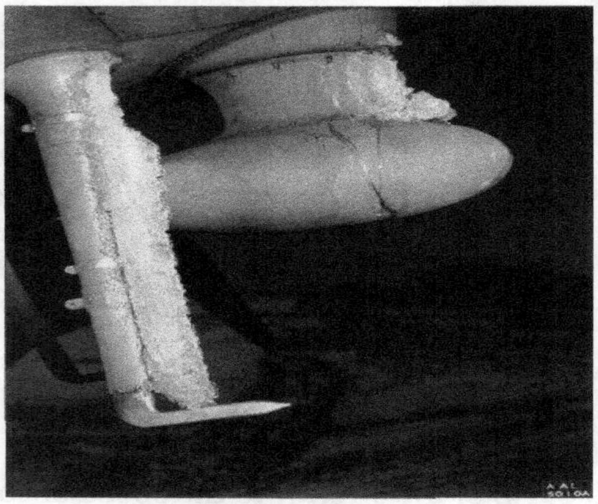

Pitot Tube Icing.

Instrument Errors

The pitot-static systems in modern aircraft are reliable, that we are always taught to "believe our instruments". However, when they do fail, the failure may be so insidious that it goes unnoticed until it's too late.

Pitot-static failures typically come in three varieties:

- Icing over the Pitot or static ports.
- Trapped water in the lines (usually after Maintenance fails to cover the ports during a wash).
- Compromise of system integrity:
  - Leaks due to holes or loose fittings.
  - Kinks in the lines.
  - Obstructions/blockages.
  - Taped or covered ports.
- Blockages in the system can cause a variety of errors.
- To prevent these errors you must complete a thorough pre-flight.

Blockages can occur from FOD, striking an object (damaging instruments), insects, trapped moisture, loss of system integrity, icing, etc.

Pitot Tube Blockage

Airspeed indicator indicates zero (gradually decreasing).

Pitot Tube and Drain-hole Blockage

Airspeed indicator will freeze and read like an altimeter as the total pressure now remains constant and the static pressure changes with climbs and descents. The measure of ram air to static air means as altitude increases and pressure decreases, the instrument will read artificially high as it is comparing it to the same dynamic (ram) pressure. Likewise, if pressure increases, such as in a descent, it will read artificially low. The aircraft will only read the correct airspeed at the altitude where the blockage occurred, assuming that static pressure is not changing.

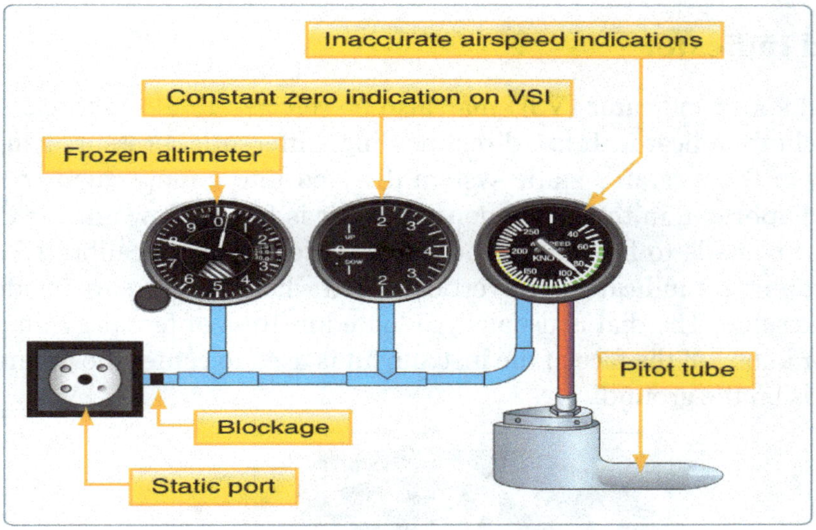

Static Blockage.

Static Blockage

If the static system becomes blocked but the pitot tube remains clear, the ASI continues to operate; however, it is inaccurate. Airspeed indicator will give erroneous readings (slower readings at altitudes above the blockage, faster below).

- The airspeed indicates lower than the actual airspeed when the aircraft is operated above the altitude where the static ports became blocked because the trapped static pressure is higher than normal for that altitude.
- When operating at a lower altitude, a faster than actual airspeed is displayed due to the relatively low static pressure trapped in the system.

If the aircraft descends, the static pressure increases on the pitot side showing an increase on the ASI. This assumes that the aircraft does not actually increase its speed.

- The increase in static pressure on the pitot side is equivalent to an increase in dynamic pressure since the pressure cannot change on the static side.

If an aircraft begins to climb after a static port becomes blocked, the airspeed begins to show a decrease as the aircraft continues to climb. This is due to the decrease in static pressure on the pitot side, while the pressure on the static side is held constant.

Some aircraft are equipped with an alternate static source in the flight deck:

- In the case of a blocked static source, opening the alternate static source introduces static pressure from the flight deck into the system.
- Flight deck static pressure is lower than outside static pressure.
- Check the aircraft AOM/POH for airspeed corrections when utilizing alternate static pressure. A blockage of the static system also affects the altimeter and VSI, as well.

Vertical Speed Indicator

An analog vertical speed indicator (VSI) may also be referred to as a vertical velocity indicator (VVI), or rate-of-climb indicator. It is a direct reading, differential pressure gauge that compares static pressure from the aircraft's static system directed into a diaphragm with static pressure surrounding the diaphragm in the instrument case. Air is free to flow unrestricted in and out of the diaphragm but is made to flow in and out of the case through a calibrated orifice. A pointer attached to the diaphragm indicates zero vertical speed when the pressure inside and outside the diaphragm are the same. The dial is usually graduated in 100s of feet per minute. A zeroing adjustment screw, or knob, on the face of the instrument is used to center the pointer exactly on zero while the aircraft is on the ground.

A typical vertical speed indicator.

As the aircraft climbs, the unrestricted air pressure in the diaphragm lowers as the air becomes less dense. The case air pressure surrounding the diaphragm lowers more slowly, having to pass through the restriction created by the orifice. This causes unequal pressure inside and outside the diaphragm, which in turn causes the diaphragm to contract a bit and the pointer, indicates a climb. The process works in reverse for an aircraft in a descent. If a steady climb or descent is maintained, a steady pressure differential is established between the diaphragm and case pressure surrounding it, resulting in an accurate indication of the rate of climb via graduations on the instrument face.

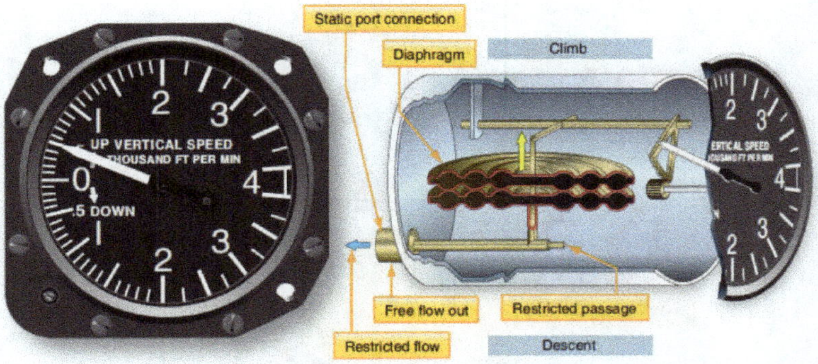

The VSI is a differential pressure gauge that compares free-flowing static air pressure in the diaphragm with restricted static air pressure around the diaphragm in the instrument case.

A shortcoming of the rate-of-climb mechanism as described is that there is a lag of six to nine seconds before a stable differential pressure can be established that indicates the actual climb or descent rate of the aircraft. An instantaneous vertical speed indicator (IVSI) has a built-in mechanism to reduce this lag. A small, lightly sprung dashpot, or piston, reacts to the direction change of an abrupt climb or descent. As this small accelerometer does so, it pumps air into or out of the diaphragm, hastening the establishment of the pressure differential that causes the appropriate indication.

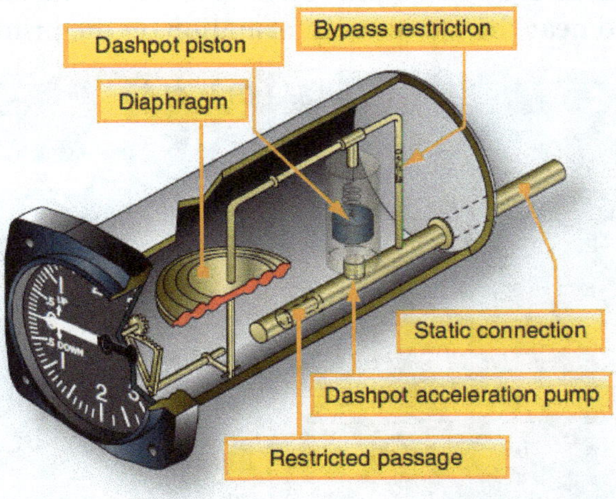

The small dashpot in this IVSI reacts abruptly to a climb or descent pumping air into or out of the diaphragm causing an instantaneously vertical speed indication.

Gliders and lighter-than-air aircraft often make use of a variometer. This is a differential VSI that compares static pressure with a known pressure. It is very sensitive and gives an instantaneous indication. It uses a rotating vane with a pointer attached to it. The vane separates two chambers. One is connected to the aircraft's static vent or is open to the atmosphere. The other is connected to a small reservoir inside the instrument that is filled to a known pressure. As static air pressure increases, the pressure in the static air chamber increases and pushes against the vane. This rotates the vane and pointer, indicating a descent since the static pressure is now greater than the set amount in the chamber with reservoir pressure. During a climb, the reservoir pressure is greater than the static pressure; the vane is pushed in the opposite direction, causing the pointer to rotate and indicate a climb.

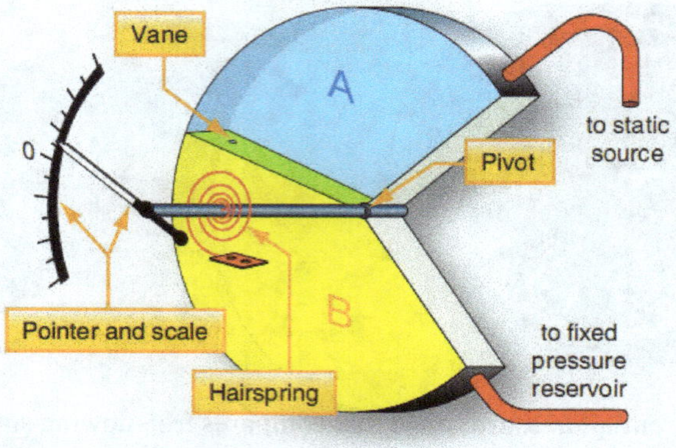

A variometer uses differential pressure to indicate vertical speed. A rotating vane separating two chambers (one with static pressure, the other with a fixed pressure reservoir), moves the pointer as static pressure changes.

The rate-of-climb indication in a digitally displayed instrument system is computed from static air input to the ADC. An aneroid, or solid-state pressure sensor, continuously reacts to changes in static pressure. The digital clock within the computer replaces the calibrated orifice found on an analog instrument. As the static pressure changes, the computer's clock can be used to develop a rate for the change. Using the known lapse rate conversion for air pressure as altitude increases or decreases, a figure for climb or descent in fpm can be calculated and sent to the cockpit. The vertical speed is often displayed near the altimeter information on the primary flight display.

This primary flight display unit of a Garmin 1000 series glass cockpit instrumentation package for light aircraft indicates altitude using a vertical linear scale and a numerical counter. As the aircraft climbs or descends, the scale behind the black numerical altitude readout changes.

Altimeter

An altimeter is a type of barometer which measures vertical distance to the surface, necessary for a pilot to maintain an assigned altitude. Altimeters function by measuring pressure differences which are displayed in terms of feet (or meters). Different types of altimeters measure different types of altitude. Altimeters however, are not without their errors. This is in part due to their construction, but also due to the effects of atmospheric pressure variations. Pilots conduct their preflight actions which include inspections, tests, and regulation to mitigate any anomalies within the altimeter. Altimeter technology has not remained static and the advent of Inertial Reference Units (IRU), Inertial Navigation Systems (INS), and Attitude Heading Reference Systems (AHRS).

Altimeter Purpose

Altimeters exist as a means of maintaining altitude. Air Traffic Control issues altimeter (pressure) settings to pilots of a local area to ensure everyone is using the same reference datum. This is especially important during Instrument Flight Rules where separation responsibilities fall upon ATC.

Unless under control of ATC, altitudes are not assigned to pilots however, to ensure safety and instrumentation accuracy, local altimeter settings are still very important.

- Instances where VFR pilots need to be accurate with their altitude include flying VFR cruising altitudes, flying traffic patterns, and maintaining minimum heights above protected areas.

Altimeter Function

The altimeter functions like a barometer, detecting pressure differences from a standard (29.92 in-Hg, aka, absolute pressure) using the static system. The altimeter can be calibrated for non-standard pressure by changing the pressure setting in the Kollsman window.

- Rotating the knob changes both the barometric scale and the altimeter pointers in such a way that a change in the barometric scale of 1" Hg changes the pointer indication by 1,000' which indicates the standard pressure lapse rate.

- The scale is set based on the local AWOS/ASOS/ATIS reading.

Function

As an aircraft increases in altitude the aneroid capsules which are calibrated to 29.92 expand due to decreasing ambient air pressure which moves the needles through gears and linkages to show an increase in altitude.

Conversely, as the aircraft descends, increasing pressure on the capsules causes them to contract while gears and linkages display a decrease in altitude.

- When set to the local altimeter setting, the altimeter shows height above sea level (true altitude), but not above such land features as hills, mountains, or valleys (absolute altitude).

- If set to the local field elevation prior to takeoff, such as the case with gliders, the altimeter will show height above the ground (absolute altitude), but not above sea level (sea level).

Types of Altimeters

Different altimeters measure altitude through various methods such as radio, radar, laser, or capacitive technology.

Drum Type

Instrument Flying Handbook, Drum-Type Altimeter.

These instruments have only one pointer that makes one revolution for every 1,000'. Each number represents 100' and each mark represents 20'. A drum, marked in thousands of feet, is geared to the mechanism that drives the pointer. To read this type of altimeter, first look at the drum to get the thousands of feet, and then at the pointer to get the feet and hundreds of feet.

Sensitive Altimeter

Instrument Flying Handbook, Three Pointer Altimeter.

One with an adjustable barometric scale allowing the pilot to set the reference pressure from which the altitude is measured. This scale is visible in a small window called the Kollsman window. A knob on the instrument adjusts the scale. The range of the scale is from 28.00" to 31.00" inches of mercury (Hg), or 948 to 1,050 millibars.

Absolute Altimeter

Also referred to as a radio altimeter, indicating the altitude over objects below based on radio signal pulses.

Types of Altitude

- Absolute altitude: The height of an aircraft above the terrain over which it is flying, usually expressed in AGL (above ground level) (HAA, HAT, TCH).

- Indicated Altitude: After setting the altimeter at the current altimeter setting, this is the uncorrected altitude read directly from that altimeter.
- True Altitude: Actual aircraft's height above sea level, usually expressed as MSL (mean sea level); all elevations on aeronautical charts are expressed in terms of true altitude.
- Pressure Altitude:
    - After adjusting the altimeter's setting to 29.92 mb or 1,013.2 mb, this is the altimeter reading that corresponds to the altitude in the standard atmosphere where the pressure is the same as you are.
    - Calculate pressure altitude.
    - LAGS (inversely proportional).
        - If the given altimeter setting is less than 29.92, then you ADD to your altitude.
        - If the given altimeter setting is greater than 29.92, then you SUBTRACT from your altitude.
- Density Altitude: The pressure altitude corrected for temperature variations. This is most important to takeoff and climb performance of an aircraft. Calculate density altitude.
- GPS Altitude: GPS Altitude is a measure take from, and therefore in reference to, the GPS Satellite Constellation. Since pressure readings passed to pilots are taken on the ground, atmospherics will guarantee a disparity between GPS altitude and pressure altitude as altitude increases. Although GPS altitude can be highly accurate, it is usually not referenced while in flight. This ensures all aircraft are using the same reference, and any error from pressure settings is generally shared across all aircraft.

Altimeter Construction

Consists of evacuated, corrugated bronze aneroid capsules which connect to needles on the face of the instrument. A striped segment is visible below 10,000', while above, a mask begins to cover it until 15,000', where it is fully covered. Stripes start to cover at 10,000' and completely cover at 15,000'.

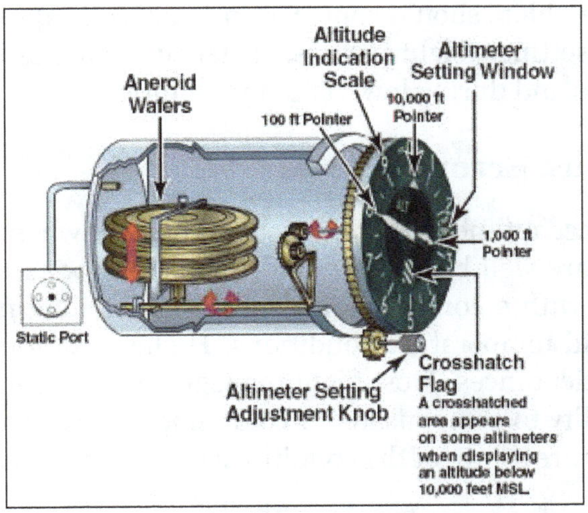

Instrument flying handbook, sensitive altimeter components.

Altimeter Errors

The accuracy of an aircraft's altimeter depends on a number of factors:

- Nonstandard atmospheric temperature.
- Nonstandard atmospheric pressure.
- Aircraft static pressure systems (position error).
- Instrument error.

Non-standard Atmospheric Temperature

The crucial values to consider are standard temperature versus the ambient (at altitude) temperature. It is this "difference" that causes the error in indicated altitude. When the column of air is warmer than standard, you are higher than your altimeter indicates. Subsequently, when the column of air is colder than standard, you are lower than indicated.

When flying into a cooler air mass while maintaining a constant indicated altitude, you are losing true altitude. However, flying into a cooler air mass does not necessarily mean you will be lower than indicated if the difference is still on the plus side. For example, while flying at 10,000 feet (where STANDARD temperature is +5 degrees Celsius (C)), the outside air temperature cools from +5 degrees C to 0 degrees C, and the temperature error will nevertheless cause the aircraft to be HIGHER than indicated. It is the extreme "cold" difference that normally would be of concern to the pilot. Also, when flying in cold conditions over mountainous terrain, the pilot should exercise caution in flight planning both in regard to route and altitude to ensure adequate en route and terminal area terrain clearance.

Non-standard temperatures can result in a change to effective vertical paths and actual descent rates while using aircraft Baro-VNAV equipment for vertical guidance on final approach segments. A higher than standard temperature will result in a steeper gradient and increased actual descent rate. Indications of these differences are often not directly related to vertical speed indications. Conversely, a lower than standard temperature will result in a shallower descent gradient and reduced actual descent rate. Pilots should consider potential consequences of these effects on approach minimums, power settings, sight picture, visual cues, etc., especially for high-altitude or terrain-challenged locations and during low-visibility conditions.

ICAO Cold Temperature Error Table

The cold temperature induced altimeter error may be significant when considering obstacle clearances when temperatures are well below standard. Pilots may wish to increase their minimum terrain clearance altitudes with a corresponding increase in ceiling from the normal minimum when flying in extreme cold temperature conditions. Higher altitudes may need to be selected when flying at low terrain clearances. Most flight management systems (FMS) with air data computers implement a capability to compensate for cold temperature errors. Pilots flying with these systems should ensure they are aware of the conditions under which the system will automatically compensate.

If compensation is applied by the FMS or manually, ATC must be informed that the aircraft is not flying the assigned altitude.

- Otherwise, vertical separation from other aircraft may be reduced, creating a potentially hazardous situation.

To use the table, find the reported temperature in the left column, and then read across the top row to the height above the airport/reporting station. Subtract the airport elevation from the altitude of the final approach fix (FAF). The intersection of the column and row is the amount of possible error.

- Example: The reported temperature is -10 °Celsius and the FAF is 500' above the airport elevation.

- The reported current altimeter setting may place the aircraft as much as 50' below the altitude indicated by the altimeter.

When using the cold temperature error table, the altitude error is proportional to both the height above the reporting station elevation and the temperature at the reporting station. For IFR approach procedures, the reporting station elevation is assumed to be airport elevation. It is important to understand that corrections are based upon the temperature at the reporting station, not the temperature observed at the aircraft's current altitude and height above the reporting station and not the charted IFR altitude.

To see how corrections are applied, note the following example:

- Airport Elevation 496'.

- Airport Temperature – 50 °C.

A charted IFR approach to the airport provides the following data:

- Minimum Procedure Turn Altitude: 1,800'.

- Minimum FAF Crossing Altitude: 1,200'.

- Straight-in Minimum Descent Altitude; 800'.

- Circling MDA: 1,000'.

The Minimum Procedure Turn Altitude of 1,800' will be used as an example to demonstrate determination of the appropriate temperature correction. Typically, altitude values are rounded up to the nearest 100' level. The charted procedure turn altitude of 1,800' minus the airport elevation of 500' equals 1,300'. The altitude difference of 1,300' falls between the correction chart elevations of 1,000' and 1,500'. At the station temperature of -50 °C, the correction falls between 300' and 450'. Dividing the difference in compensation values by the difference in altitude above the airport gives the error value per foot.

In this case, 150' divided by 500' = 0.33' for each additional foot of altitude above 1,000'. This provides a correction of 300' for the first 1,000' and an additional value of 0.33 times 300', or 99', which is rounded to 100'. 300' + 100' = total temperature correction of 400'. For the given conditions, correcting the charted value of 1,800' above MSL (equal to a height above the reporting station of 1,300') requires the addition of 400'. Thus, when flying at an indicated altitude of 2,200', the aircraft is actually flying a true altitude of 1,800'.

Minimum Procedure Turn Altitude 1,800' charted = 2,200' corrected. Minimum FAF Crossing Altitude 1,200' charted = 1,500' corrected. Straight-in MDA 800' charted = 900' corrected. Circling MDA 1,000' charted = 1,200' corrected.

Reported Temp 0 °C	Height Above Airport in Feet													
	200	300	400	500	600	700	800	900	1000	1500	2000	3000	4000	5000
+10	10	10	10	10	20	20	20	20	20	30	40	60	80	90
0	20	20	30	30	40	40	50	50	60	90	120	170	230	280
-10	20	30	40	50	60	70	80	90	100	150	200	290	390	490
-20	30	50	60	70	90	100	120	130	140	210	280	420	570	710
-30	40	60	80	100	120	140	150	170	190	280	380	570	760	950
-40	50	80	100	120	150	170	190	220	240	360	480	720	970	1210
-50	60	90	120	150	180	210	240	270	300	450	590	890	1190	1500

Instrument flying handbook, ICAO cold temperature error table.

Non-standard Atmospheric Pressure

Once in flight, it is very important to obtain frequently current altimeter settings en route. If you do not reset your altimeter when flying from an area of high pressure into an area of low pressure, your aircraft will be closer to the surface than your altimeter indicates. An inch error in the altimeter setting equals 1,000 feet of altitude. To quote an old saying "going from a high to a low, look out below".

TBL 7-2-3, derived from ICAO formulas, indicates how much error can exist when the temperature is extremely cold. To use the table, find the reported temperature in the left column, then read across the top row to locate the height above the airport/reporting station (i.e., subtract the airport/ reporting elevation from the intended flight altitude). The intersection of the column and row is how much lower the aircraft may actually be as a result of the possible cold temperature induced error.

Position Error

Most aircraft do not have the capability of setting an altimeter above 31.00 in-Hg or below 28.00 in-Hg.

- For aircraft with the capability of setting the current altimeter setting and operating into airports with the capability of measuring the current altimeter setting, no additional restrictions apply.

The possible result of the above example should be obvious, particularly if operating at the minimum altitude or when conducting an instrument approach. When operating in extreme cold temperatures, pilots may wish to compensate for the reduction in terrain clearance by adding a cold temperature correction. A sensitive altimeter is designed to indicate standard changes from standard conditions, but most flying involves errors caused by non-standard conditions and the pilot must be able to modify the indications to correct for these errors.

Instrument Errors

Table: Lowest usable flight level.

Altimeter Setting (Current Reported)	Lowest Usable Flight Level
29.92 or higher	180
29.91 to 29.42	185
29.41 to 28.92	190
28.91 to 28.42	195
28.41 to 27.92	200

Most pressure altimeters are subject to mechanical, elastic, temperature, and installation errors. Although manufacturing and installation specifications, as well as the periodic test and inspections required by regulations (14 CFR Part 43, Appendix E), act to reduce these errors, any scale error may be observed in the following manner:

- Set the current reported altimeter setting on the altimeter setting scale.
- Altimeter should now read field elevation if you are located on the same reference level used to establish the altimeter setting.
- Note the variation between the known field elevation and the altimeter indication. If this variation is in the order of plus or minus 75 feet, the accuracy of the altimeter is questionable and the problem should be referred to an appropriately rated repair station for evaluation and possible correction.

Once in flight, it is very important to obtain frequently current altimeter settings en route. If you do not reset your altimeter when flying from an area of high pressure into an area of low pressure, your aircraft will be closer to the surface than your altimeter indicates. An inch error in the altimeter setting equals 1,000 feet of altitude. To quote an old saying "going from a high to a low, look out below".

TBL 7-2-3, derived from ICAO formulas, indicates how much error can exist when the temperature is extremely cold. To use the table, find the reported temperature in the left column, then read across the top row to locate the height above the airport/reporting station (i.e., subtract the airport/reporting elevation from the intended flight altitude). The intersection of the column and row is how much lower the aircraft may actually be as a result of the possible cold temperature induced error.

Pilots are responsible to compensate for cold temperature altimetry errors when operating into an airport with any published cold temperature restriction and a reported airport temperature at or below the published temperature restriction. Pilots must ensure compensating aircraft are correcting on the proper segment or segments of the approach. Manually correct if compensating aircraft system is inoperable. Pilots manually correcting, are responsible to calculate and apply a cold temperature altitude correction derived from TBL 7−2−3 to the affected approach segment or segments. Pilots must advise the cold temperature altitude correction to Air Traffic Control (ATC). Pilots are not required to advise ATC of a cold temperature altitude correction inside of the final approach fix.

The possible result of the above example should be obvious, particularly if operating at the minimum altitude or when conducting an instrument approach. When operating in extreme cold temperatures, pilots may wish to compensate for the reduction in terrain clearance by adding a cold temperature correction.

- Example: Temperature-10 degrees Celsius and the aircraft altitude is 1,000 feet above the airport elevation. The chart shows that the reported current altimeter setting may place the aircraft as much as 100 feet below the altitude indicated by the altimeter.

Inherent Altimeter Errors

When the aircraft is flying in air that is warmer than standard, the air is less dense and the pressure levels are farther apart.

- When the aircraft is flying at an indicated altitude of 5,000', the pressure level for that altitude is higher than it would be in air at standard temperature, and the aircraft is higher than it would be if the air were cooler.

If the air is colder than standard, it is denser and the pressure levels are closer together.

- When the aircraft is flying at an indicated altitude of 5,000', its true altitude is lower than it would be if the air were warmer. Most pressure altimeters are subject to mechanical, elastic, temperature, and installation errors.

Mechanical Errors

Errors with the instrument itself. If the indication checked during preflight is off by more than 75' from the surveyed elevation, the instrument should be referred to a certificated instrument repair station for recalibration. A difference between ambient temperature and pressure causes an erroneous indication on the altimeter.

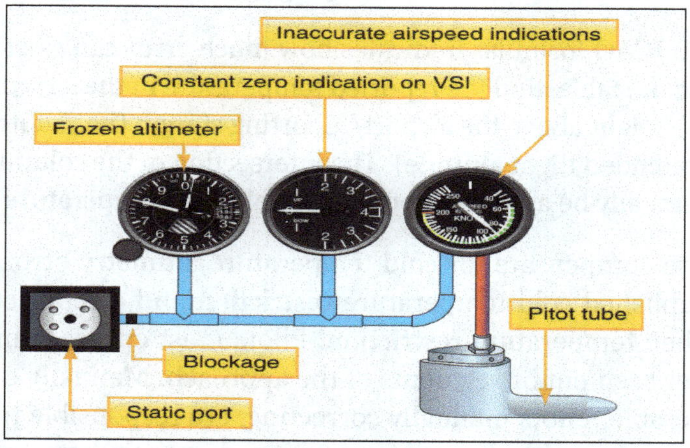

Static Blockage.

Static Port Blockage

Trapped static pressure causes the altimeter to freeze at the altitude where the blockage occurred.

Cold Weather Altimeter Errors

A correctly calibrated pressure altimeter indicates true altitude above mean sea level (MSL) when operating within the International Standard Atmosphere (ISA) parameters of pressure and temperature. Nonstandard pressure conditions are corrected by applying the correct local area altimeter setting.

Temperature errors from ISA result in true altitude being higher than indicated altitude whenever the temperature is warmer than ISA and true altitude being lower than indicated altitude whenever the temperature is colder than ISA.

True altitude variance under conditions of colder than ISA temperatures poses the risk of inadequate obstacle clearance. Under extremely cold conditions, pilots may need to add an appropriate temperature correction determined from the chart in figure to charted IFR altitudes to ensure terrain and obstacle clearance with the following restrictions:

- Altitudes specifically assigned by Air Traffic Control (ATC), such as "maintain 5,000" shall not be corrected. Assigned altitudes may be rejected if the pilot decides that low temperatures pose a risk of inadequate terrain or obstacle clearance.
- If temperature corrections are applied to charted IFR altitudes (such as procedure turn altitudes, final approach fix crossing altitudes, etc.), the pilot must advise ATC of the applied correction.

Table: Lowest flight level correction factor.

Altimeter Setting	Correction Factor
29.92 or higher	None
29.91 to 29.42	500 feet
29.41 to 28.92	1000 feet
28.91 to 28.42	1500 feet
28.41 to 27.92	2000 feet
27.91 to 27.42	2500 feet

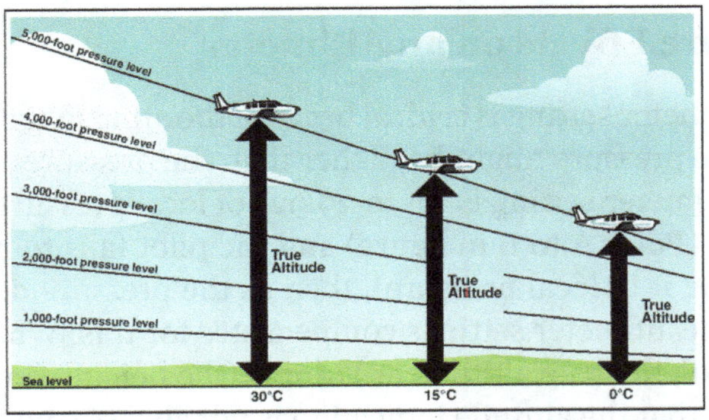

Instrument Flying Handbook, The loss of altitude experienced when flying into an area where the air is colder (more dense) than standard.

Altimeter Enhancements

It is not sufficient in the airspace system for only the pilot to have an indication of the aircraft's altitude; the air traffic controller on the ground must also know the altitude of the aircraft. To provide this information, the aircraft is typically equipped with an encoding altimeter.

When the ATC transponder is set to Mode C, the encoding altimeter supplies the transponder with a series of pulses identifying the flight level (in increments of 100') at which the aircraft is flying. This series of pulses is transmitted to the ground radar where they appear on the controller's scope as an alphanumeric display around the return for the aircraft.

The transponder allows the ground controller to identify the aircraft and determine the pressure altitude at which it is flying. A computer inside the encoding altimeter measures the pressure referenced from 29.92" Hg and delivers this data to the transponder. When the pilot adjusts the barometric scale to the local altimeter setting, the data sent to the transponder is not affected. This is to ensure that all Mode C aircraft are transmitting data referenced to a common pressure level.

ATC equipment adjusts the displayed altitudes to compensate for local pressure differences allowing display of targets at correct altitudes. 14 CFR part 91 requires the altitude transmitted by the transponder to be within 125' of the altitude indicated on the instrument used to maintain flight altitude.

Preflight Action

Should indicate within 75' of field elevation when set with the proper barometric pressure (not regulatory) according to Aeronautical Information Manual 7-2-3. Operating with a known altimeter error, especially in IFR conditions, may not be illegal but certainly falls under careless and reckless operation.

If you set your altimeter according to ATIS/AWOS and it is not perfect for field elevation, remember your location on the field. Filed elevation the highest elevation of any usable landing surface. Gliders may set the airport elevation so they can know AGL. Obtaining an altimeter setting:

- Altimeter settings can be obtained from ATC, automated broadcasts, weather reports, as well as various other applications and programs.

Nonstandard Pressure Effects on an Altimeter

Maintaining a current altimeter setting is critical because the atmosphere pressure is not constant. That is, in one location the pressure might be higher than the pressure just a short distance away. Take an aircraft whose altimeter setting is set to 29.92" of local pressure. As the aircraft moves to an area of lower pressure (Point A to B in figure) and the pilot fails to readjust the altimeter setting (essentially calibrating it to local pressure), then as the pressure decreases, the true altitude will be lower. Adjusting the altimeter settings compensates for this. When the altimeter shows an indicated altitude of 5,000', the true altitude at Point A (the height above mean sea level) is only 3,500' at Point B. The fact that the altitude indication is not always true lends itself to the memory aid, "When flying from hot to cold or from a high to a low.

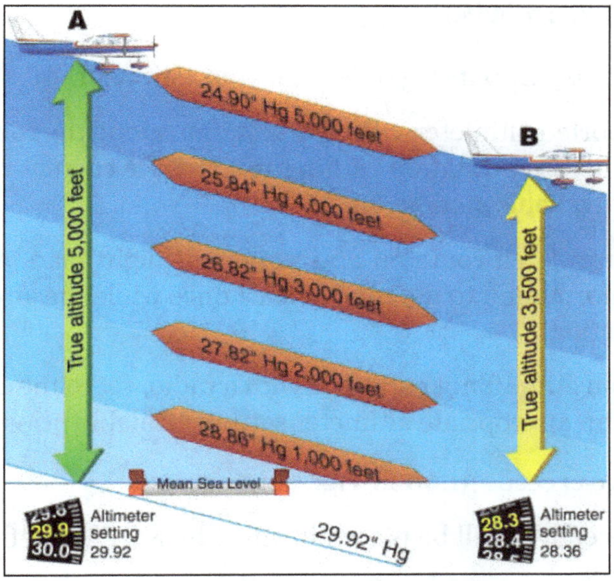

Instrument flying handbook, effects of nonstandard pressure on an altimeter of an aircraft flown into air of lower than standard pressure (air is less dense).

High Barometric Pressure

Cold, dry air masses may produce barometric pressures in excess of 31.00 inches of Mercury, and many altimeters do not have an accurate means of being adjusted for settings of these levels.

In this case, the aircraft actual altitude will be higher than the altimeter indicates. When the barometric pressure exceeds 31.00 inches, air traffic controllers will issue the actual altimeter setting, and:

- En Route/Arrivals Advise pilots to remain set on 31.00 inches until reaching the final approach segment.
- Departures Advise pilots to set 31.00 inches prior to reaching any mandatory/crossing altitude or 1,500 feet, whichever is lower.
- The altimeter error caused by the high pressure will be in the opposite direction to the error caused by the cold temperature.

Low Barometric Pressure

When abnormally low barometric pressure conditions occur (below 28.00), flight operations by aircraft unable to set the actual altimeter setting are not recommended. This is different from high barometric pressure because the true altitude of the aircraft is lower than the indicated altitude if the pilot is unable to set the actual altimeter setting.

Altimeter Procedures

The cruising altitude or flight level of aircraft must be maintained by reference to an altimeter. Altimeter setting procedures vary based on altitude and atmospheric pressures.

Operations below 18,000' MSL

When barometric pressure is 31.00 in-Hg or less set your altimeter to:

- To the current reported altimeter setting of a station along the route and within 100 NM of the aircraft, or if there is no station within this area, the current reported altimeter setting of an appropriate available station.
    - When an aircraft is en route on an instrument flight plan, air traffic controllers will furnish this information to the pilot at least once while the aircraft is in the controller's area of jurisdiction.
    - In the case of an aircraft not equipped with a radio, set to the elevation of the departure airport or use an appropriate altimeter setting available prior to departure.

When barometric pressure exceeds 31.00 in-Hg:

- The following procedures will be placed in effect by NOTAM defining the geographic area affected:
    - For all aircraft Set 31.00 inches for en route operations below 18,000 feet MSL.
    - Maintain this setting until beyond the affected area or until reaching final approach segment.
    - At the beginning of the final approach segment, the current altimeter setting will be set, if possible.
    - If not possible, 31.00 inches will remain set throughout the approach.
    - Aircraft on departure or missed approach will set 31.00 inches prior to reaching any mandatory/crossing altitude or 1,500' AGL, whichever is lower.
        - Air traffic control will issue actual altimeter settings and advise pilots to set 31.00 inches in their altimeters for en route operations below 18,000 feet MSL in affected areas.
- During preflight, barometric altimeters must be checked for normal operation to the extent possible.
- For aircraft with the capability of setting the current altimeter setting and operating into airports with the capability of measuring the current altimeter setting, no additional restrictions apply.
- For aircraft operating VFR, there are no additional restrictions; however, extra diligence in flight planning and in operating in these conditions is essential.
- Airports unable to accurately measure barometric pressures above 31.00 inches of Hg. will report the barometric pressure as "missing" or "in excess of 31.00 inches of Hg." Flight operations to and from those airports are restricted to VFR weather conditions.
- When the altimeter cannot be set to the higher pressure setting, the aircraft actual altitude will be higher than the altimeter indicates.

- For aircraft operating IFR and unable to set the current altimeter setting, the following restrictions apply:
    - To determine the suitability of departure alternate airports, destination airports, and destination alternate airports, increase ceiling requirements by 100 feet and visibility requirements by 1/4 statute mile for each 1/10 of an inch of Hg., or any portion thereof, over 31.00 inches. These adjusted values are then applied in accordance with the requirements of the applicable operating regulations and operations specifications.
    - Example: Destination altimeter is 31.28 inches, ILS DH 250 feet (200-1/2). When flight planning, add 300-3/4 to the weather requirements which would become 500-11/4.
    - On approach, 31.00 inches will remain set. Decision height (DH) or minimum descent altitude must be deemed to have been reached when the published altitude is displayed on the altimeter.
    - Although visibility is normally the limiting factor on an approach, pilots should be aware that when reaching DH the aircraft will be higher than indicated. Using the example above the aircraft would be approximately 300 feet higher.
    - These restrictions do not apply to authorized Category II and III ILS operations nor do they apply to certificate holders using approved QFE altimetry systems.

Table: Lowest usable flight level.

Altimeter	Lowest FL
29.92 or higher	FL 180
29.91 to 29.42	FL 185
29.41 to 28.92	FL 190
28.91 to 28.42	FL 195
28.41 to 27.92	FL 200
27.91 to 27.42	FL 205
27.41 to 26.92	FL 210

Low Barometric Pressure (Below 28.00 in-Hg)

Flight operations by aircraft unable to set the actual altimeter setting are not recommended.

- The true altitude of the aircraft is lower than the indicated altitude if the pilot is unable to set the actual altimeter setting.

Operations at or above 18,000' MSL

Pilots set the altimeter to the standard 29.92 when entering the "flight levels". Since VFR pilots can fly at 17,500' MSL, atmospheric pressures below 29.92 in Hg (think "when flying high to low, lookout below") will bring those pilots flying at 18,000' MSL lower, decreasing the required 500' separation from possible VFR traffic.

As a result, ATC establishes a lowest usable flight level depending upon the atmospheric pressure in an area of operation. The lowest usable flight level is determined by the atmospheric pressure in the area of operation as shown in figure.

- Example: The minimum safe altitude of a route is 19,000' MSL and the altimeter setting is reported between 29.92 and 29.42 inches of mercury, the lowest usable flight level will be 195, which is the flight level equivalent of 19,500 feet MSL (minimum altitude plus 500 feet).

Table: Altimeter adjustment factor.

Altimeter Setting	Correction Factor
29.92 or higher	None
29.91 to 29.42	500 feet
29.41 to 28.92	1000 feet
28.91 to 28.42	1500 feet
28.41 to 27.92	2000 feet
27.91 to 27.42	2500 feet

Inertial Reference Unit, Inertial Navigation System and Attitude Heading Reference System

IRUs are self-contained systems comprised of gyros and accelerometers that provide aircraft attitude (pitch, roll, and heading), position, and velocity information in response to signals resulting from inertial effects on system components.

- Once aligned with a known position, IRUs continuously calculate position and velocity. IRU position accuracy decays with time.
- This degradation is known as "drift".
- INSs combine the components of an IRU with an internal navigation computer. By programming a series of waypoints, these systems will navigate along a predetermined track.

AHRSs are electronic devices that provide attitude information to aircraft systems such as weather radar and autopilot, but do not directly compute position information. Aircraft equipped with slaved compass systems may be susceptible to heading errors caused by exposure to magnetic field disturbances (flux fields) found in materials that are commonly located on the surface or buried under taxiways and ramps.

- These materials generate a magnetic flux field that can be sensed by the aircraft's compass system flux detector or "gate", which can cause the aircraft's system to align with the material's magnetic field rather than the earth's natural magnetic field.
- The system's erroneous heading may not self-correct.
- Prior to take off pilots should be aware that a heading misalignment may have occurred during taxi.

- Pilots are encouraged to follow the manufacturers or other appropriate procedures to correct possible heading misalignment before takeoff is commenced.

Machmeter

A machmeter is an instrument which provides an indication of the Mach Number, (M), which is the ratio between the aircraft true air speed (TAS) and the local speed of sound (LSS). This ratio, which equals one when the TAS is equal to the local speed of sound, is very important in aircraft operating at high speed.

The machmeter uses the aircraft pitot-static system to generate M and usually portrays this on a simple needle and dial instrument, such as that shown in figure.

Alternatively, the machmeter may be combined with the Air Speed Indicator (ASI), in which case it is often referred to as a Combined Speed Indicator (CSI).

High speed aircraft, including airliners and business jets, have limiting mach numbers which must not be deliberately exceeded. If the aircraft is deliberately or accidentally allowed to exceed its limiting mach, shock waves are likely to form on the aerofoils and can result in buffet or mach tuck.

Some aircraft use a constant mach number (rather than constant speed) technique for cruise operations. Constant mach technique may be used to separate aircraft on the same track and at the same altitude whilst in a non-radar environment.

Compass Systems

A compass system indicates the bearing, or angle of the direction in which an aircraft is pointing in the horizontal plane. A compass may indicate magnetic heading or bearing, bearing referenced to a radio signal source, or bearing with respect to an inertially maintained line of position.

Magnetic Compass

One of the oldest and simplest instruments for indicating direction is the magnetic compass. It is also one of the basic instruments required by Title 14 of the Code of Federal Regulations (14 CFR) part 91 for both VFR and IFR flight.

A magnet is a piece of material, usually a metal containing iron, which attracts and holds lines of magnetic flux. Regardless of size, every magnet has two poles north and south. When one magnet is placed in the field of another, the unlike poles attract each other, and like poles repel.

An aircraft magnetic compass, such as the one in figure, has two small magnets attached to a metal float sealed inside a bowl of clear compass fluid similar to kerosene. A graduated scale, called a card, is wrapped around the float and viewed through a glass window with a lubber line across it. The card is marked with letters representing the cardinal directions, north, east, south, and west, and a number for each 30° between these letters. The final "0" is omitted from these directions. For example, 3 = 30°, 6 = 60°, and 33 = 330°. There are long and short graduation marks between the letters and numbers, each long mark representing 10° and each short mark representing 5°.

Driven by signals from a flux valve, the compass card in this RMI indicates the heading of the aircraft opposite the upper center index mark. The green pointer is driven by the ADF.

The float and card assembly has a hardened steel pivot in its center that rides inside a special, spring-loaded, hard glass jewel cup. The buoyancy of the float takes most of the weight off the pivot, and the fluid damps the oscillation of the float and card. This jewel-and-pivot type mounting allows the float freedom to rotate and tilt up to approximately 18° angle of bank. At steeper bank angles, the compass indications are erratic and unpredictable.

The compass housing is entirely full of compass fluid. To prevent damage or leakage when the fluid expands and contracts with temperature changes, the rear of the compass case is sealed with a flexible diaphragm, or with a metal bellows in some compasses.

The magnets align with the Earth's magnetic field and the pilot reads the direction on the scale opposite the lubber line. Note that in figure, the pilot sees the compass card from its backside. When the pilot is flying north as the compass shows, east is to the pilot's right. On the card, "33", which represents 330° (west of north), is to the right of north. The reason for this apparent backward graduation is that the card remains stationary, and the compass housing and the pilot turn around it, always viewing the card from its backside.

A magnetic compass: The vertical line is called the lubber line.

A compensator assembly mounted on the top or bottom of the compass allows an aviation maintenance technician (AMT) to create a magnetic field inside the compass housing that cancels the influence of local outside magnetic fields. This is done to correct for deviation error. The compensator assembly has two shafts whose ends have screwdriver slots accessible from the front of the compass. Each shaft rotates one or two small compensating magnets. The end of one shaft is marked E-W, and its magnets affect the compass when the aircraft is pointed east or west. The other shaft is marked N-S and its magnets affect the compass when the aircraft is pointed north or south.

Vertical Card Magnetic Compass

The floating magnet type of compass not only has all the errors just described, but also lends itself to confused reading. It is easy to begin a turn in the wrong direction because its card appears backward. East is on what the pilot would expect to be the west side. The vertical card magnetic compass eliminates some of the errors and confusion. The dial of this compass is graduated with letters representing the cardinal directions, numbers every 30°, and tick marks every 5°. The dial is rotated by a set of gears from the shaft-mounted magnet, and the nose of the symbolic aircraft on the instrument glass represents the lubber line for reading the heading of the aircraft from the dial. Eddy currents induced into an aluminum-damping cup damp, or decrease, oscillation of the magnet.

Vertical card compass.

Lags or Leads

When starting a turn from a northerly heading, the compass lags behind the turn. When starting a turn from a southerly heading, the compass leads the turn.

Eddy Current Damping

The decreased amplitude of oscillations by the interaction of magnetic fields. In the case of a vertical card magnetic compass, flux from the oscillating permanent magnet produces eddy currents in a damping disk or cup. The magnetic flux produced by the eddy currents opposes the flux from the permanent magnet and decreases the oscillations.

Gyromagnetic Compass

The gyromagnetic compass is a horizontal gyro slaved to magnetic north by a magnetic field detector (flux valve) usually located in the wing tips of the aircraft away from metal structure. The short term accuracy of the gyro combined with the long term accuracy of the magnetic compass provides a stable and accurate heading reference.

Gyromagnetic Compass System

A Gyromagnetic compass system comprises:

- A magnetic detector (flux valve) which electronically senses the horizontal component of the earth's magnetic field.
- A turn/accelerator cut out switch which cuts out the feed from the flux valve to the gyro during turns and acceleration.
- A gyroscope which maintains its position reference and continues to point in the same direction regardless of what the aircraft does.
- Compass repeaters to transmit heading information to the associated instruments.

Operating Modes

The GMC has two operational modes (selectable by the pilot through the GMC control panel):

- Magnetic slaved information (also revered to as "SLAVE" mode) – this mode is used when magnetic heading references are reliable. In this mode, the directional gyro is slaved to the magnetic flux detector, which supplies long-term magnetic reference to correct the apparent drift of the gyro. In order to ensure correct indication of the directional gyro indicator it is necessary to align it with the magnetic compass at regular intervals, approximately every 10 to 15 minutes. On some modern aircraft this alignment is performed automatically by the control system.

- Directional gyro (also referred to as "FREE" mode) – this mode is used when magnetic heading references are not reliable (e.g. in polar regions). In this mode, the system furnishes an inertial heading reference, with corrections introduced manually to offset earth rate and other errors.

Magnetic Compass Induced Errors

Variation

The Earth rotates about its geographic axis; maps and charts are drawn using meridians of longitude that pass through the geographic poles. Directions measured from the geographic poles are called true directions. The magnetic North Pole to which the magnetic compass point is not collocated with the geographic North Pole, but is some 1,300 miles away; directions measured from the magnetic poles are called magnetic directions. In aerial navigation, the difference between true and magnetic directions is called variation. This same angular difference in surveying and land navigation is called declination.

Figure shows the isogonic lines that identify the number of degrees of variation in their area. The line that passes near Chicago is called the agonic line. Anywhere along this line the two poles are aligned, and there is no variation. East of this line, the magnetic North Pole is to the west of the geographic North Pole and a correction must be applied to a compass indication to get a true direction.

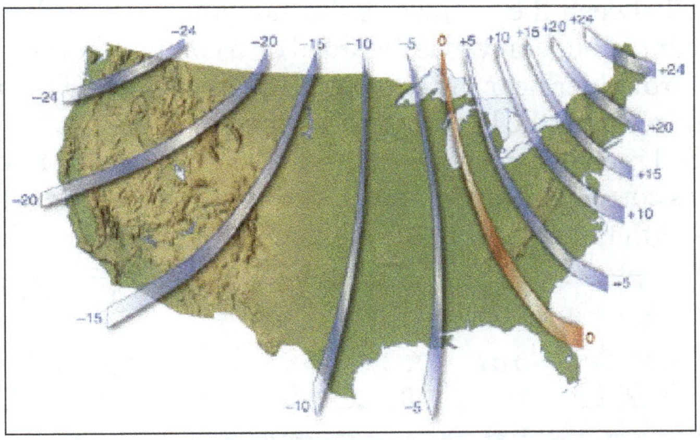

Isogonic lines are lines of equal variation.

Flying in the Washington, D.C., area, for example, the variation is 10° west. If a pilot wants to fly a true course of south (180°), the variation must be added to this, resulting in a magnetic course of 190° to fly. Flying in the Los Angeles, California, area, the variation is 14° east. To fly a true course of 180° there, the pilot would have to subtract the variation and fly a magnetic course of 166°. The variation error does not change with the heading of the aircraft; it is the same anywhere along the isogonic line.

Deviation

The magnets in a compass align with any magnetic field. Local magnetic fields in an aircraft caused by electrical current flowing in the structure, in nearby wiring or any magnetized part of the structure, conflict with the Earth's magnetic field and cause a compass error called deviation.

Deviation, unlike variation, is different on each heading, but it is not affected by the geographic location. Variation error cannot be reduced or changed, but deviation error can be minimized when an AMT performs the maintenance task known as "swinging the compass."

Most airports have a compass rose, which is a series of lines marked out on a ramp or maintenance run-up area where there is no magnetic interference. Lines, oriented to magnetic north, are painted every 30°, as shown in figure.

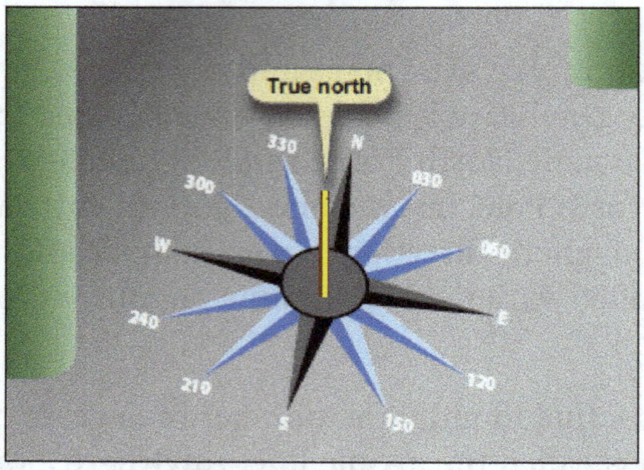

Utilization of a compass rose aids compensation for deviation errors.

The AMT aligns the aircraft on each magnetic heading and adjusts the compensating magnets to minimize the difference between the compass indication and the actual magnetic heading of the aircraft. Any error that cannot be removed is recorded on a compass correction card, like the one in figure, and placed in a cardholder near the compass. The pilot can taxi the aircraft to the compass rose and maneuver the aircraft to the headings prescribed by the AMT, and if authorized to do so, the AMT can also taxi and maneuver the aircraft; however, only the AMT can adjust the compass or complete the compass correction card. If the pilot wants to fly a magnetic heading of 120° and the aircraft is operating with the radios on, the pilot should fly a compass heading of 123°.

FOR	000	030	060	090	120	150
STEER						
RDO. ON	001	032	062	095	123	155
RDO. OFF	002	031	064	094	125	157

FOR	180	210	240	270	300	330
STEER						
RDO. ON	176	210	243	271	296	325
RDO. OFF	174	210	240	273	298	327

A compass correction card shows the deviation correction for any heading.

The corrections for variation and deviation must be applied in the correct sequence and is shown below, starting from the true course desired.

Determine the Magnetic Course

True Course (180°) ± Variation (+10°) = Magnetic Course (190°). The magnetic course (190°) is steered if there is no deviation error to be applied. The compass card must now be considered for the compass course of 190°.

Determine the Compass Course

Magnetic Course (190°, from step 1) ± Deviation (−2°, from correction card) = Compass Course (188°). Intermediate magnetic courses between those listed on the compass card need to be interpreted. Therefore, to steer a true course of 180°, the pilot would follow a compass course of 188°.

To find the true course that is being flown when the compass course is known:

Compass Course ± Deviation = Magnetic Course ± Variation= True Course.

Dip Errors

The lines of magnetic flux are considered to leave the Earth at the magnetic North Pole and enter at the magnetic South Pole. At both locations the lines are perpendicular to the Earth's surface. At the magnetic equator, which is halfway between the poles, the lines are parallel with the surface. The magnets in a compass align with this field, and near the poles they dip, or tilt, the float and card. The float is balanced with a small dip-compensating weight, to dampen the effects of dip when operating in the middle latitudes of the northern hemisphere. This dip (and weight) causes two very noticeable errors northerly turning error and acceleration error.

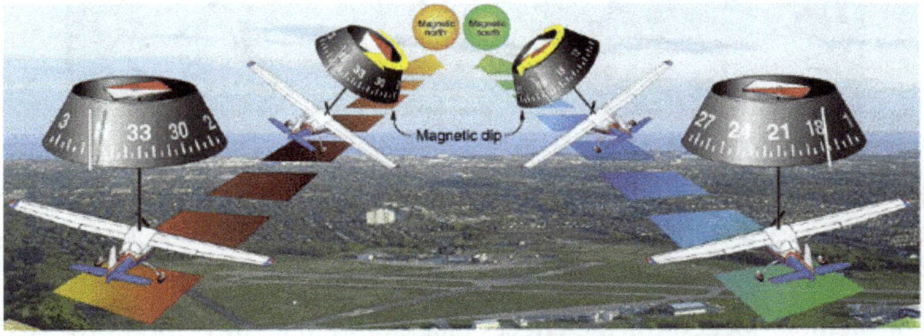

Northerly turning error.

The pull of the vertical component of the Earth's magnetic field causes northerly turning error, which is apparent on a heading of north or south. When an aircraft flying on a heading of north makes a turn toward east, the aircraft banks to the right, and the compass card tilts to the right. The vertical component of the Earth's magnetic field pulls the north seeking end of the magnet to the right, and the float rotates, causing the card to rotate toward west, the direction opposite the direction the turn is being made.

If the turn is made from north to west, the aircraft banks to the left and the compass card tilts down on the left side. The magnetic field pulls on the end of the magnet that causes the card to rotate toward east. This indication is again opposite to the direction the turn is being made. The rule for this error is when starting a turn from a northerly heading, the compass indication lags behind the turn.

When an aircraft is flying on a heading of south and begins a turn toward east, the Earth's magnetic field pulls on the end of the magnet that rotates the card toward east, the same direction the turn is being made. If the turn is made from south toward west, the magnetic pull starts the card rotating toward west—again, in the same direction the turn is being made. The rule for this error is when starting a turn from a southerly heading; the compass indication leads the turn.

In acceleration error, the dip-correction weight causes the end of the float and card marked N (the south-seeking end) to be heavier than the opposite end. When the aircraft is flying at a constant speed on a heading of east or west, the float and card is level. The effects of magnetic dip and the weight are approximately equal. If the aircraft accelerates on a heading of east, the inertia of the weight holds its end of the float back and the card rotates toward north. As soon as the speed of the aircraft stabilizes, the card swings back to its east indication. If, while flying on this easterly heading, the aircraft decelerates, the inertia causes the weight to move ahead and the card rotates toward south until the speed again stabilizes.

Effects of acceleration error.

When flying on a heading of west, the same things happen. Inertia from acceleration causes the weight to lag, and the card rotates toward north. When the aircraft decelerates on a heading of west, inertia causes the weight to move ahead and the card rotates toward south.

A mnemonic, or memory jogger, for the effect of acceleration error is the word "ANDS" (acceleration—north, deceleration— south). Acceleration causes an indication toward north; deceleration causes an indication toward south.

Oscillation Error

Oscillation is a combination of all of the other errors, and it results in the compass card swinging back and forth around the heading being flown. When setting the gyroscopic heading indicator to agree with the magnetic compass, use the average indication between the swings.

Gyroscopic Flight Instruments

The instrument containing gyroscopes are the turn coordinator, heading indicator, and the attitude indicator. To understand how these instruments operate requires knowledge of the instrument power systems, gyroscopic principles, and the operating principles of each instrument.

Gyroscopic Principles

Any spinning object exhibits gyroscopic properties. A wheel or rotor designed and mounted to utilize these properties is called a gyroscope. Two important design characteristics of an instrument gyro are great weight for its size, or high density, and rotation at high speed with low friction bearings.

There are two general types of mountings; the type used depends upon which property of the gyro is utilized. A freely or universally mounted gyroscope is free to rotate in any direction about its center of gravity. Such a wheel is said to have three planes of freedom. The wheel or rotor is free to rotate in any plane in relation to the base and is balanced so that, with the gyro wheel at rest, it remains in the position in which it is placed. Restricted or semi-rigidly mounted gyroscopes are those mounted so that one of the planes of freedom is held fixed in relation to the base. There are two fundamental properties of gyroscopic action rigidity in space and precession.

Rigidity in Space

Rigidity in space refers to the principle that a gyroscope remains in a fixed position in the plane in which it is spinning. An example of rigidity in space is that of a bicycle wheel. As the bicycle wheels increase speed, they become more stable in their plane of rotation. This is why a bicycle is unstable and maneuverable at low speeds and stable and less maneuverable at higher speeds.

By mounting this wheel, or gyroscope, on a set of gimbal rings, the gyro is able to rotate freely in any direction. Thus, if the gimbal rings are tilted, twisted, or otherwise moved, the gyro remains in the plane in which it was originally spinning.

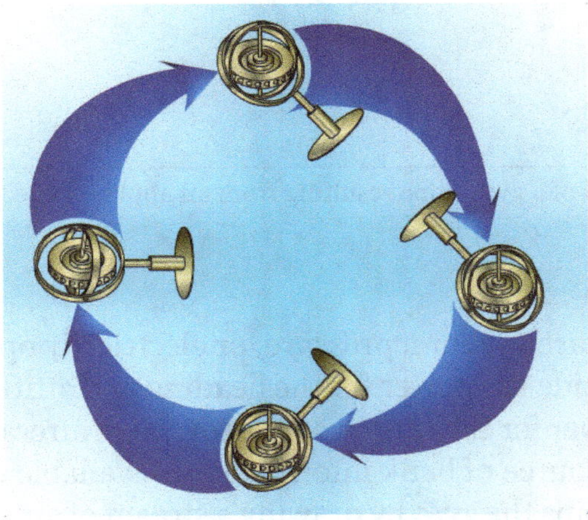

Regardless of the position of its base, a gyro tends to remain rigid in space, with its axis of rotation pointed in a constant direction.

Precession

Precession is the tilting or turning of a gyro in response to a deflective force. The reaction to this force does not occur at the point at which it was applied; rather, it occurs at a point that is 90° later in the direction of rotation. This principle allows the gyro to determine a rate of turn by sensing the amount of pressure created by a change in direction. The rate at which the gyro processes is inversely proportional to the speed of the rotor and proportional to the deflective force.

Using the example of the bicycle, precession acts on the wheels in order to allow the bicycle to turn. While riding at normal speed, it is not necessary to turn the handle bars in the direction of the desired turn. A rider simply leans in the direction that he or she wishes to go. Since the wheels are rotating in a clockwise direction when viewed from the right side of the bicycle, if a rider leans to the left, a force is applied to the top of the wheel to the left. The force actually acts 90° in the direction of rotation, which has the effect of applying a force to the front of the tire, causing the bicycle to move to the left. There is a need to turn the handlebars at low speeds because of the instability of the slowly turning gyros and also to increase the rate of turn.

Precession can also create some minor errors in some instruments. Precession can cause a freely spinning gyro to become displaced from its intended plane of rotation through bearing friction, etc. Certain instruments may require corrective realignment during flight, such as the leading indicator.

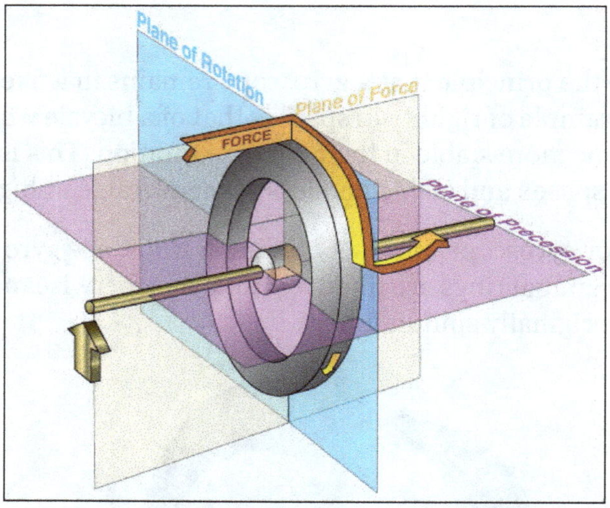

Precession of a gyroscope resulting from an applied deflective force.

Sources of Power

In some aircraft, all the gyros are vacuum, pressure, or electrically operated. In other aircraft, vacuum or pressure systems provide the power for the heading and attitude indicators, while the electrical system provides the power for the turn coordinator. Most aircraft have at least two sources of power to ensure at least one source of bank information is available if one power source fails. The vacuum or pressure system spins the gyro by drawing a stream of air against the rotor vanes to spin the rotor at high speed, much like the operation of a waterwheel or turbine. The amount of vacuum or pressure required for instrument operation varies, but is usually between 4.5 "Hg and 5.5 "Hg.

One source of vacuum for the gyros is a vane-type engine driven pump that is mounted on the accessory case of the engine. Pump capacity varies in different aircraft, depending on the number of gyros.

A typical vacuum system consists of an engine-driven vacuum pump, relief valve, air filter, gauge, and tubing necessary to complete the connections. The gauge is mounted in the aircraft's instrument panel and indicates the amount of pressure in the system (vacuum is measured in inches of mercury less than ambient pressure).

As shown in figure, air is drawn into the vacuum system by the engine-driven vacuum pump. It first goes through a filter, which prevents foreign matter from entering the vacuum or pressure system. The air then moves through the attitude and heading indicators where it causes the gyros to spin. A relief valve prevents the vacuum pressure, or suction, from exceeding prescribed limits.

After that, the air is expelled overboard or used in other systems, such as for inflating pneumatic deicing boots.

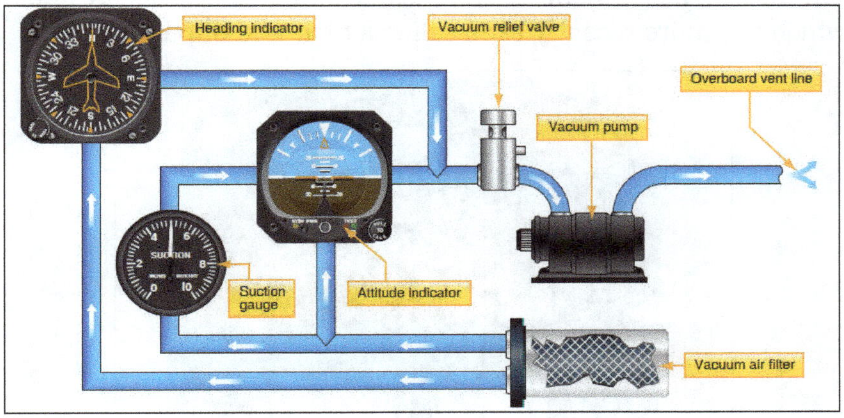

Typical vacuum system.

It is important to monitor vacuum pressure during flight, because the attitude and heading indicators may not provide reliable information when suction pressure is low. The vacuum, or suction, gauge is generally marked to indicate the normal range. Some aircraft are equipped with a warning light that illuminates when the vacuum pressure drops below the acceptable level.

When the vacuum pressure drops below the normal operating range, the gyroscopic instruments may become unstable and inaccurate. Cross-checking the instrument routinely is a good habit to develop.

Turn Indicator

Turn Indicator is an instrument for indicating either the amount or the rate of turn of an airplane about its vertical axis. Aircraft use two types of turn indicators turn-and-slip indicator and turn coordinator. Because of the way the gyro is mounted, the turn-and-slip indicator shows only the rate of turn in degrees per second. The turn coordinator is mounted at an angle, or canted, so it can initially show roll rate. When the roll stabilizes, it indicates rate of turn. Both instruments indicate turn direction and quality (coordination), and also serve as a backup source of bank information in the event an attitude indicator fails. Coordination is achieved by referring to the inclinometer, which consists of a liquid-filled curved tube with a ball inside.

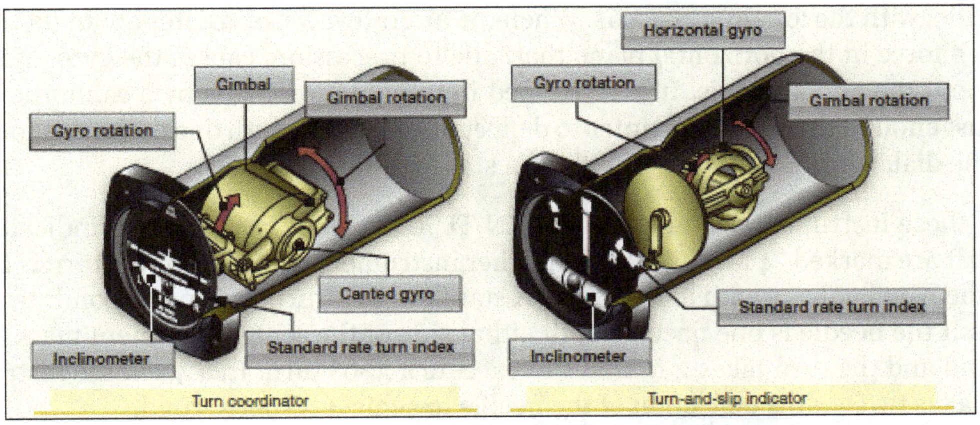

Turn indicators rely on controlled precession for their operation.

Turn-and-slip Indicator

The first gyroscopic aircraft instrument was the turn indicator in the needle and ball, or turn-and-bank indicator, which has more recently been called a turn-and-slip indicator.

Turn-and-slip indicator.

The inclinometer in the instrument is a black glass ball sealed inside a curved glass tube that is partially filled with a liquid for damping. This ball measures the relative strength of the force of gravity and the force of inertia caused by a turn. When the aircraft is flying straight-and-level, there is no inertia acting on the ball, and it remains in the center of the tube between two wires. In a turn made with a bank angle that is too steep, the force of gravity is greater than the inertia and the ball rolls down to the inside of the turn. If the turn is made with too shallow a bank angle, the inertia is greater than gravity and the ball rolls upward to the outside of the turn.

The inclinometer does not indicate the amount of bank, nor does it indicate slip; it only indicates the relationship between the angle of bank and the rate of yaw.

The turn indicator is a small gyro spun either by air or by an electric motor. The gyro is mounted in a single gimbal with its spin axis parallel to the lateral axis of the aircraft and the axis of the gimbal parallel with the longitudinal axis. When the aircraft yaws, or rotates about its vertical axis, it produces a force in the horizontal plane that, due to precession, causes the gyro and its gimbal to rotate about the gimbal's axis. It is restrained in this rotation plane by a calibration spring; it rolls over just enough to cause the pointer to deflect until it aligns with one of the doghouse-shaped marks on the dial, when the aircraft is making a standard rate turn.

The dial of these instruments is marked "2 MIN TURN." Some turn-and-slip indicators used in faster aircraft are marked "4 MIN TURN." In either instrument, a standard rate turn is being made whenever the needle aligns with a doghouse. A standard rate turn is 3° per second. In a 2 minute instrument, if the needle is one needle width either side of the center alignment mark, the turn is 3° per second and the turn takes 2 minutes to execute a 360° turn. In a 4 minute instrument, the same turn takes two widths deflection of the needle to achieve 3° per second.

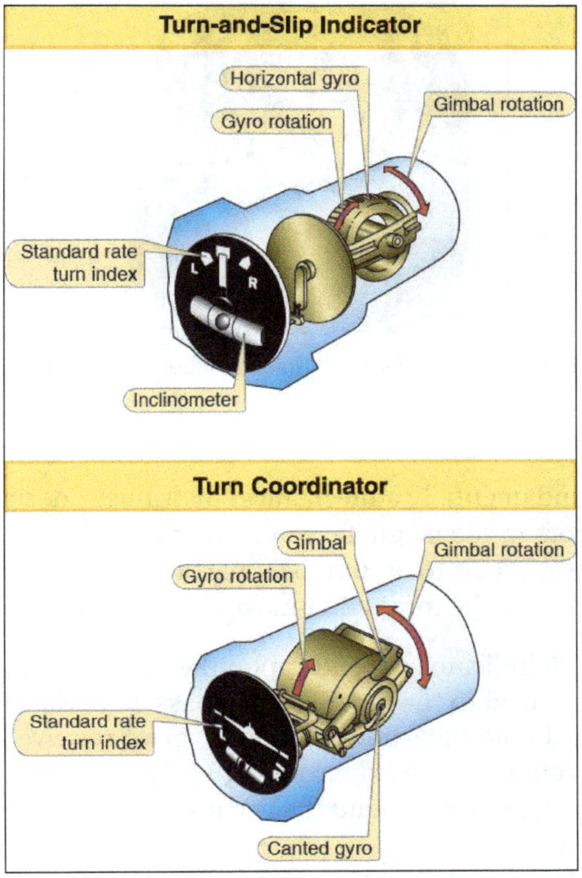

The rate gyro in both turn-and-slip indicator and turn coordinator.

Turn Coordinator

The major limitation of the older turn-and-slip indicator is that it senses rotation only about the vertical axis of the aircraft. It tells nothing of the rotation around the longitudinal axis, which in normal flight occurs before the aircraft begins to turn. A turn coordinator operates on precession, the same as the turn indicator, but its gimbals frame is angled upward about 30° from the longitudinal axis of the aircraft. This allows it to sense both roll and yaw. Therefore during a turn, the indicator first shows the rate of banking and once stabilized the turn rate. Some turn coordinator gyros are dual-powered and can be driven by either air or electricity.

Rather than using a needle as an indicator, the gimbal moves a dial that is the rear view of a symbolic aircraft. The bezel of the instrument is marked to show wings-level flight and bank angles for a standard rate turn.

The inclinometer, similar to the one in a turn-and-slip indicator, is called a coordination ball, which shows the relationship between the bank angle and the rate of yaw. The turn is coordinated when the ball is in the center, between the marks. The aircraft is skidding when the ball rolls toward the outside of the turn and is slipping when it moves toward the inside of the turn. A turn coordinator does not sense pitch. This is indicated on some instruments by placing the words "no pitch information" on the dial.

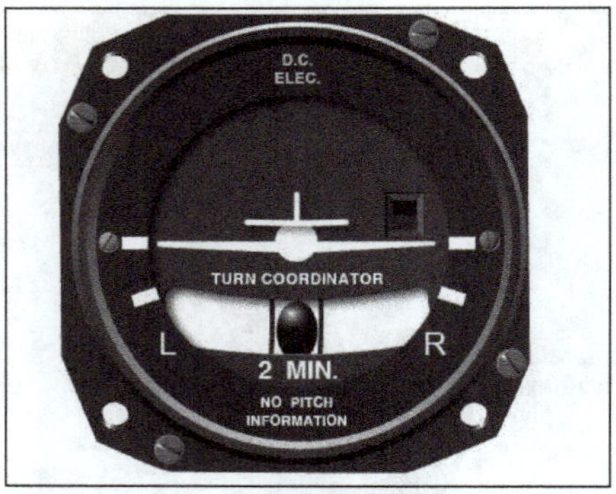

A turn coordinator senses rotation about both roll and yaw axes.

Heading Indicator

The heading indicator is fundamentally a mechanical instrument designed to facilitate the use of the magnetic compass. Errors in the magnetic compass are numerous, making straight flight and precision turns to headings difficult to accomplish, particularly in turbulent air. A heading indicator, however, is not affected by the forces that make the magnetic compass difficult to interpret.

The operation of the heading indicator depends upon the principle of rigidity in space. The rotor turns in a vertical plane and fixed to the rotor is a compass card. Since the rotor remains rigid in space, the points on the card hold the same position in space relative to the vertical plane of the gyro. The aircraft actually rotates around the rotating gyro, not the other way around. As the instrument case and the aircraft revolve around the vertical axis of the gyro, the card provides clear and accurate heading information.

Because of precession caused by friction, the leading indicator creeps or drifts from a heading to which it is set. Among other factors, the amount of drift depends largely upon the condition of the instrument. If the bearings are worn, dirty, or improperly lubricated, the drift may be excessive. Another error in the heading indicator is caused by the fact that the gyro is oriented in space, and the Earth rotates in space at a rate of 15° in 1 hour. Thus, discounting precession caused by friction, the leading indicator may indicate as much as 15° error per every hour of operation.

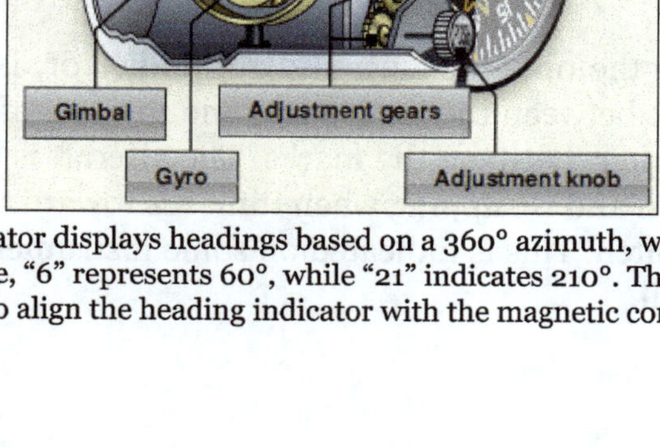

A heading indicator displays headings based on a 360° azimuth, with the final zero omitted. For example, "6" represents 60°, while "21" indicates 210°. The adjustment knob is used to align the heading indicator with the magnetic compass.

Some heading indicators referred to as horizontal situation indicators (HSI) receive a magnetic north reference from a magnetic slaving transmitter, and generally need no adjustment. The magnetic slaving transmitter is called a magnetometer.

Attitude and Heading Reference System

Electronic flight displays have replaced free-spinning gyros with solid-state laser systems that are capable of flight at any attitude without tumbling. This capability is the result of the development of the Attitude and Heading Reference System (AHRS).

The AHRS sends attitude information to the PFD in order to generate the pitch and bank information of the attitude indicator. The heading information is derived from a magnetometer which senses the earth's lines of magnetic flux. This information is then processed and sent out to the PFD to generate the heading display.

Attitude and heading reference system (AHRS).

Navigational Systems

Very High Frequency Omnidirectional Range (VOR)

VOR is the primary navigational aid (NAVAID) used by civil aviation in the National Airspace System (NAS). The VOR ground station is oriented to magnetic north and transmits azimuth information to the aircraft, providing 360 courses TO or FROM the VOR station. When DME is installed with the VOR, it is referred to as a VOR/DME and provides both azimuth and distance information. When military tactical air navigation (TACAN) equipment is installed with the VOR, it is known as a VORTAC and provides both azimuth and distance information.

The courses oriented FROM the station are called radials. The VOR information received by an aircraft is not influenced by aircraft attitude or heading. Radials can be envisioned to be like the spokes of a wheel on which the aircraft is on one specific radial at any time. For example, aircraft A (heading 180°) is inbound on the 360° radial; after crossing the station, the aircraft is outbound on the 180° radial at A1. Aircraft B is shown crossing the 225° radial. Similarly, at any point around the station, an aircraft can be located somewhere on a specific VOR radial. Additionally, a VOR needle on an RMI will always point to the course that will take you to the VOR station where conversely the ADF needle points to the station as a RB from the aircraft. In the example above, the ADF needle at position A would be pointed straight ahead, at A1 to the aircraft's 180° position (tail) and at B, to the aircraft's right.

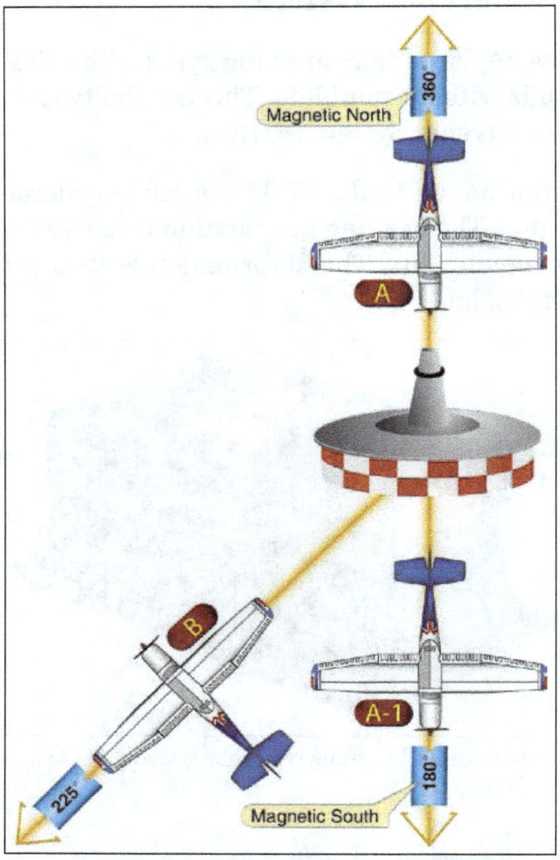

VOR Radials: The VOR receiver measures and presents information to indicate bearing TO or FROM the station. In addition to the navigation signals transmitted by the VOR, a Morse code signal is transmitted concurrently to identify the facility, as well as voice transmissions for communication and relay of weather and other information.

VORs are classified according to their operational uses. The standard VOR facility has a power output of approximately 200 watts, with a maximum usable range depending upon the aircraft altitude, class of facility, location of the facility, terrain conditions within the usable area of the facility, and other factors. Above and beyond certain altitude and distance limits, signal interference from other VOR facilities and a weak signal make it unreliable. Coverage is typically at least 40 miles at normal minimum instrument flight rules (IFR) altitudes. VORs with accuracy problems in parts of their service volume are listed in Notices to Airmen (NOTAMs) and in the Airport/Facility Directory (A/FD) under the name of the NAVAID.

VOR Components

The ground equipment consists of a VOR ground station, which is a small, low building topped with a fl at white disc, upon which are located the VOR antennas and a fiberglass cone-shaped tower. The station includes an automatic monitoring system. The monitor automatically turns off defective equipment and turns on the standby transmitter. Generally, the accuracy of the signal from the ground station is within 1°.

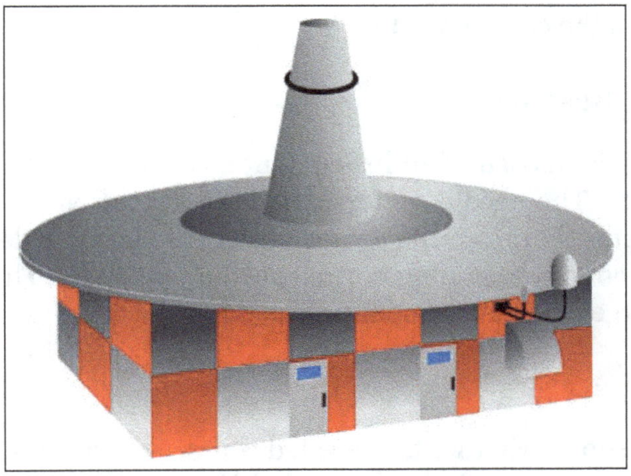

VOR Transmitter (Ground Station).

VOR facilities are aurally identified by Morse code, or voice, or both. The VOR can be used for ground-to-air communication without interference with the navigation signal. VOR facilities operate within the 108.0 to 117.95 MHz frequency band and assignment between 108.0 and 112.0 MHz is in even-tenth increments to preclude any conflict with ILS localizer frequency assignment, which uses the odd tenths in this range.

The airborne equipment includes an antenna, a receiver, and the indicator instrument. The receiver has a frequency knob to select any of the frequencies between 108.0 to 117.95 MHz. The On/Off/ volume control turns on the navigation receiver and controls the audio volume. The volume has no effect on the operation of the receiver. You should listen to the station identifier before relying on the instrument for navigation.

VOR indicator instruments have at least the essential components shown in the instrument illustrated in figure.

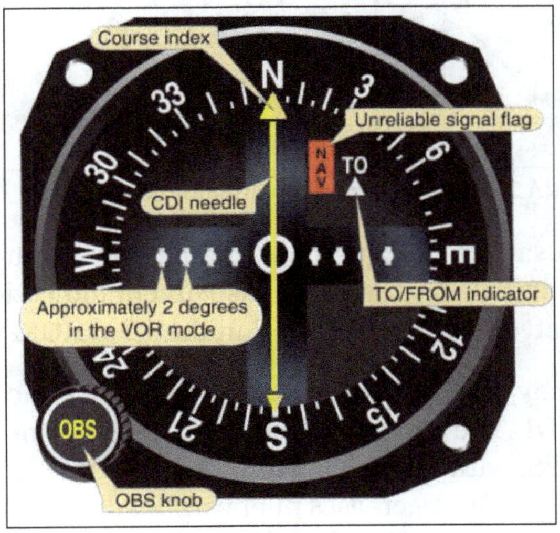

The VOR Indicator Instrument.

Omnibearing Selector

The desired course is selected by turning the OBS knob until the course is aligned with the course index mark or displayed in the course window.

Course Deviation Indicator

The deviation indicator is composed of an instrument face and a needle hinged to move laterally across the instrument face. The needle centers when the aircraft is on the selected radial or it's reciprocal. Full needle deflection from the center position to either side of the dial indicates the aircraft is 12° or more off course, assuming normal needle sensitivity. The outer edge of the center circle is 2° off course; with each dot representing an additional 2°.

TO/FROM Indicator

The TO/FROM indicator shows whether the selected course will take the aircraft TO or FROM the station. It does not indicate whether the aircraft is heading to or from the station.

Flags or other Signal Strength Indicators

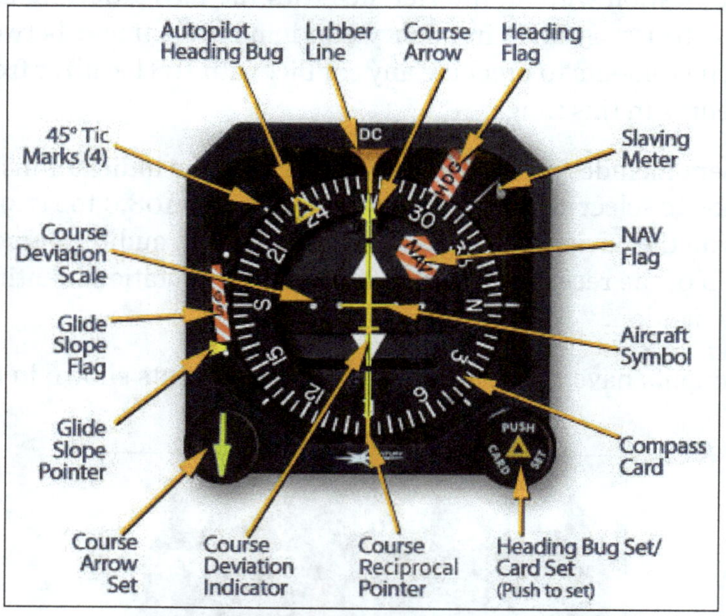

A Typical Horizontal Situation Indicator (HSI).

The device that indicates a usable or an unreliable signal may be an "OFF" flag. It retracts from view when signal strength is sufficient for reliable instrument indications. Alternately, insufficient signal strength may be indicated by a blank or OFF in the TO/FROM window.

The indicator instrument may also be a horizontal situation indicator (HSI) which combines the heading indicator and CDI. The combination of navigation information from VOR/Localizer (LOC) or from LORAN or GPS, with aircraft heading information provides a visual picture of the aircraft's location and direction. This decreases pilot workload especially with tasks such as course intercepts, flying a back-course approach, or holding pattern entry.

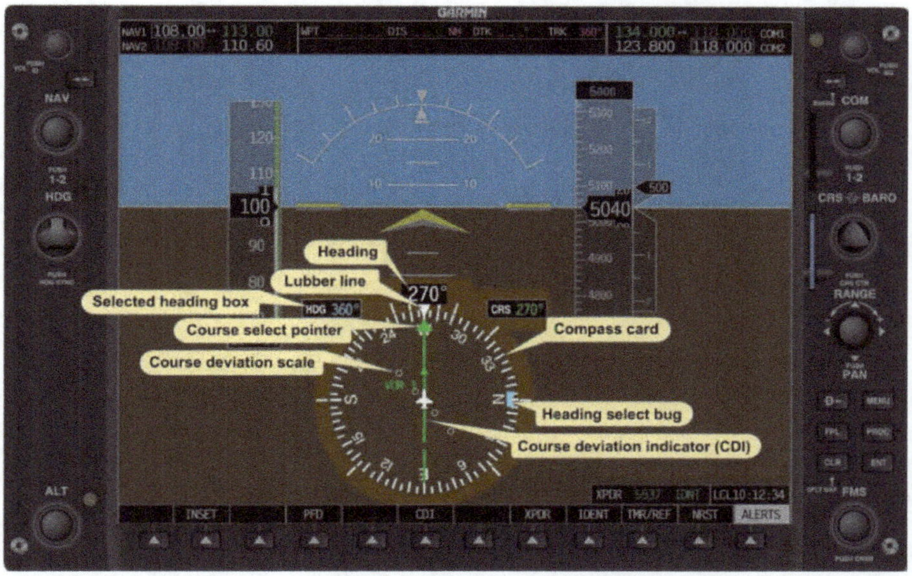

An HSI display as seen on the pilot's primary flight display (PFD) on an electronic flight instrument. Note that only attributes related to the HSI are labeled.

Function of VOR

Orientation

The VOR does not account for the aircraft heading. It only relays the aircraft direction from the station and will have the same indications regardless of which way the nose is pointing. Tune the VOR receiver to the appropriate frequency of the selected VOR ground station, turn up the audio volume, and identify the station's signal audibly. Then, rotate the OBS to center the CDI needle and read the course under or over the index.

In figure, 360° TO is the course indicated, while in figure, 180° TO is the course. The latter indicates that the aircraft (which may be heading in any direction) is, at this moment, located at any point on the 360° radial (line from the station) except directly over the station or very close to it, as between points I and S in figure. The CDI will deviate from side to side as the aircraft passes over or nearly over the station because of the volume of space above the station where the zone of confusion exists. This zone of confusion is caused by lack of adequate signal directly above the station due to the radiation pattern of the station's antenna, and because the resultant of the opposing reference and variable signals is small and constantly changing.

The CDI in figure indicates 180°, meaning that the aircraft is on the 180° or the 360° radial of the station. The TO/FROM indicator resolves the ambiguity. If the TO indicator is showing, then it is 180° TO the station. The FROM indication indicates the radial of the station the aircraft is presently on. Movement of the CDI from center, if it occurs at a relatively constant rate, indicates the aircraft is moving or drifting off the 180°/360° line. If the movement is rapid or fluctuating, this is an indication of impending station passage (the aircraft is near the station). To determine the aircraft's position relative to the station, rotate the OBS until FROM appears in the window, and then center the CDI needle. The index indicates the VOR radial where the aircraft is located. The inbound (to the station) course is the reciprocal of the radial.

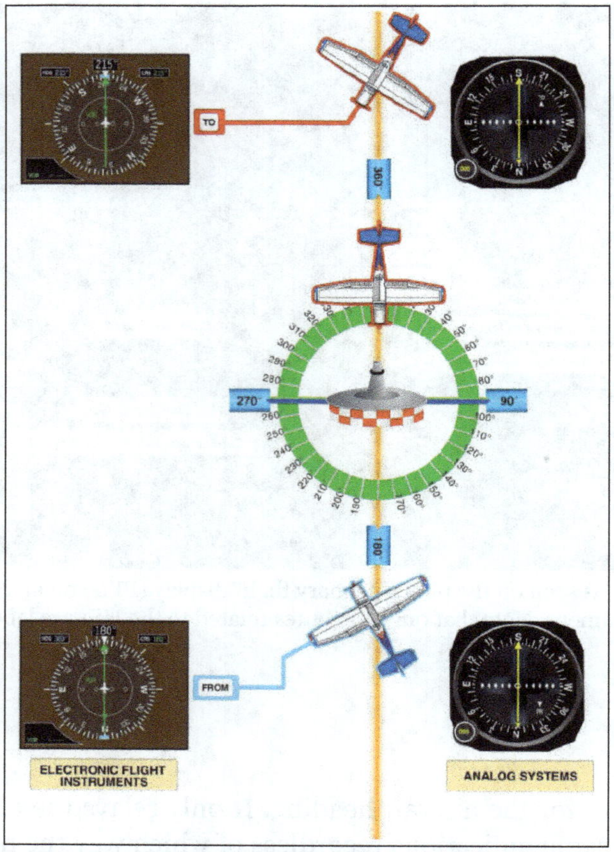

CDI Interpretation. The CDI as typically found on analog systems (right) and as found on electronic flight instruments (left).

If the VOR is set to the reciprocal of the intended course, the CDI will reflect reverse sensing. To correct for needle deflection, turn away from the needle. To avoid this reverse sensing situation, set the VOR to agree with the intended course.

A single NAVAID will allow a pilot to determine the aircraft's position relative to a radial. Indications from a second NAVAID are needed in order to narrow the aircraft's position down to an exact location on this radial.

Tracking TO and FROM the Station

To track to the station, rotate the OBS until TO appears, then center the CDI. Fly the course indicated by the index. If the CDI moves off center to the left, follow the needle by correcting course to the left, beginning with a 20° correction.

When flying the course indicated on the index, a left deflection of the needle indicates a crosswind component from the left. If the amount of correction brings the needle back to center, decrease the left course correction by half. If the CDI moves left or right now, it should do so much more slowly, and smaller heading corrections can be made for the next iteration.

Keeping the CDI centered will take the aircraft to the station. To track to the station, the OBS value at the index is not changed. To home to the station, the CDI needle is periodically centered, and the new course under the index is used for the aircraft heading. Homing will follow a circuitous route to the station, just as with ADF homing.

To track FROM the station on a VOR radial, you should first orient the aircraft's location with respect to the station and the desired outbound track by centering the CDI needle with a FROM indication. The track is intercepted by either flying over the station or establishing an intercept heading. The magnetic course of the desired radial is entered under the index using the OBS and the intercept heading held until the CDI centers. Then the procedure for tracking to the station is used to fl y outbound on the specified radial.

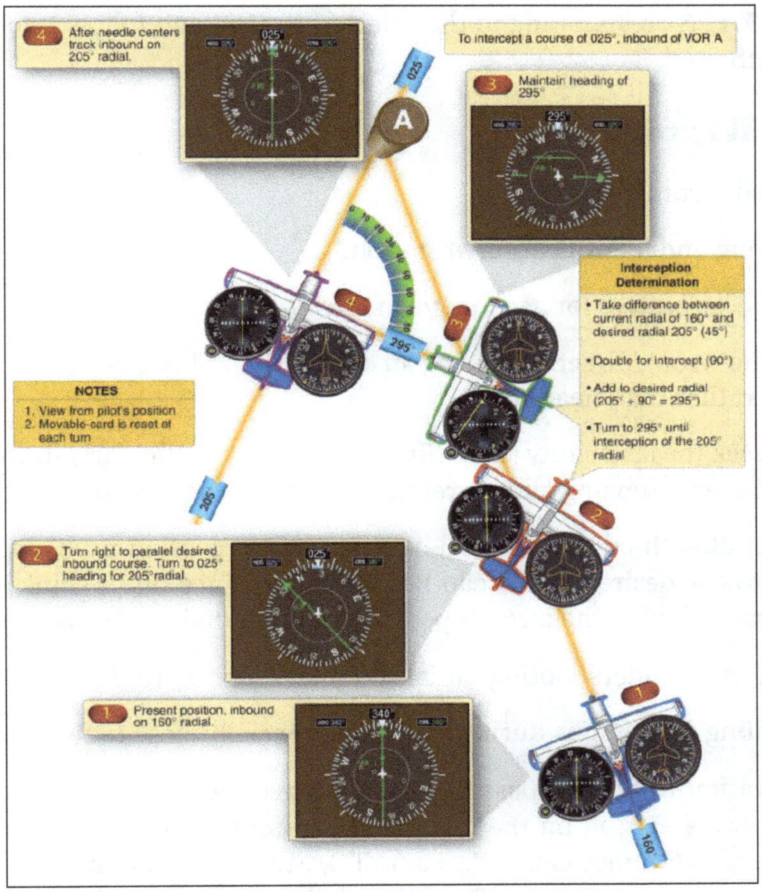

Course Interception (VOR).

Course Interception If the desired course is not the one being flown, first orient the aircraft's position with respect to the VOR station and the course to be flown, and then establish an intercept heading. The following steps may be used to intercept a predetermined course, either inbound or outbound. Steps 1–3 may be omitted when turning directly to intercept the course without initially turning to parallel the desired course.

- Turn to a heading to parallel the desired course, in the same direction as the course to be flown.
- Determine the difference between the radial to be intercepted and the radial on which the aircraft is located (205° – 160° = 045°).

- Double the difference to determine the interception angle, which will not be less than 20° nor greater than 90° (45° x 2 = 090°). 205° + 090° = 295° for the intercept).
- Rotate the OBS to the desired radial or inbound course.
- Turn to the interception heading.
- Hold this heading constant until the CDI center, which indicates the aircraft is on course. (With practice in judging the varying rates of closure with the course centerline, pilots learn to lead the turn to prevent overshooting the course).
- Turn to the MH corresponding to the selected course, and follow tracking procedures inbound or outbound.

VOR Operational Errors

Typical pilot-induced errors include:

- Careless tuning and identification of station.
- Failure to check receiver for accuracy/sensitivity.
- Turning in the wrong direction during an orientation. This error is common until visualizing position rather than heading.
- Failure to check the ambiguity (TO/FROM) indicator, particularly during course reversals, resulting in reverse sensing and corrections in the wrong direction.
- Failure to parallel the desired radial on a track interception problem. Without this step, orientation to the desired radial can be confusing. Since pilots think in terms of left and right of course, aligning the aircraft position to the radial/course is essential.
- Overshooting and undershooting radials on interception problems.
- Over controlling corrections during tracking, especially close to the station.
- Misinterpretation of station passage. On VOR receivers not equipped with an ON/OFF flag, a voice transmission on the combined communication and navigation radio (NAV/COM) in use for VOR may cause the same TO/FROM fluctuations on the ambiguity meter as shown during station passage. Read the whole receiver—TO/FROM, CDI, and OBS—before you make a decision. Do not utilize a VOR reading observed while transmitting.
- Chasing the CDI, resulting in homing instead of tracking. Careless heading control and failure to bracket wind corrections make this error common.

VOR Accuracy

The effectiveness of the VOR depends upon proper use and adjustment of both ground and airborne equipment. The accuracy of course alignment of the VOR is generally plus or minus 1°. On some VORs, minor course roughness may be observed, evidenced by course needle or brief flag alarm. At a few stations, usually in mountainous terrain, the pilot may occasionally observe a brief course needle oscillation, similar to the indication of "approaching station." Pilots flying over unfamiliar routes are cautioned to be on the alert for these vagaries, and in particular, to use the TO/ FROM indicator to determine positive station passage.

Certain propeller revolutions per minute (RPM) settings or helicopter rotor speeds can cause the VOR CDI to fluctuate as much as plus or minus 6°. Slight changes to the RPM setting will normally smooth out this roughness. Pilots are urged to check for this modulation phenomenon prior to reporting a VOR station or aircraft equipment for unsatisfactory operation.

VOR Receiver Accuracy Check

VOR system course sensitivity may be checked by noting the number of degrees of change as the OBS is rotated to move the CDI from center to the last dot on either side. The course selected should not exceed 10° or 12° either side. In addition, Title 14 of the Code of Federal Regulations (14 CFR) part 91 provides for certain VOR equipment accuracy checks, and an appropriate endorsement, within 30 days prior to flight under IFR. To comply with this requirement and to ensure satisfactory operation of the airborne system, use the following means for checking VOR receiver accuracy:

- VOR test facility (VOT) or a radiated test signal from an appropriately rated radio repair station.
- Certified checkpoints on the airport surface.
- Certified airborne checkpoints.

VOR Test Facility (VOT) The Federal Aviation Administration (FAA) VOT transmits a test signal which provides users a convenient means to determine the operational status and accuracy of a VOR receiver while on the ground where a VOT is located. Locations of VOTs are published in the A/FD. Two means of identification are used. One is a series of dots and the other is a continuous tone. Information concerning an individual test signal can be obtained from the local flight service station (FSS.) The airborne use of VOT is permitted; however, its use is strictly limited to those areas/altitudes specifically authorized in the A/FD or appropriate supplement.

To use the VOT service, tune in the VOT frequency 108.0 MHz on the VOR receiver. With the CDI centered, the OBS should read 0° with the TO/FROM indication showing FROM or the OBS should read 180° with the TO/FROM indication showing TO. Should the VOR receiver operate an RMI, it would indicate 180° on any OBS setting.

A radiated VOT from an appropriately rated radio repair station serves the same purpose as an FAA VOT signal, and the check is made in much the same manner as a VOT with some differences.

The frequency normally approved by the Federal Communications Commission (FCC) is 108.0 MHz; however, repair stations are not permitted to radiate the VOR test signal continuously. The owner or operator of the aircraft must make arrangements with the repair station to have the test signal transmitted. A representative of the repair station must make an entry into the aircraft logbook or other permanent record certifying to the radial accuracy and the date of transmission.

Certified Checkpoints

Airborne and ground checkpoints consist of certified radials that should be received at specific points on the airport surface or over specific landmarks while airborne in the immediate vicinity of the airport. Locations of these checkpoints are published in the A/FD.

Should an error in excess of ±4° be indicated through use of a ground check, or ±6° using the airborne check, IFR flight shall not be attempted without first correcting the source of the error. No correction other than the correction card figures supplied by the manufacturer should be applied in making these VOR receiver checks.

If a dual system VOR (units independent of each other except for the antenna) is installed in the aircraft, one system may be checked against the other. Turn both systems to the same VOR ground facility and note the indicated bearing to that station. The maximum permissible variation between the two indicated bearings is 4°.

Nondirectional Radio Beacon

The nondirectional radio beacon (NDB) is a ground-based radio transmitter that transmits radio energy in all directions. The ADF, when used with an NDB, determines the bearing from the aircraft to the transmitting station. The indicator may be mounted in a separate instrument in the aircraft panel. The ADF needle points to the NDB ground station to determine the relative bearing (RB) to the transmitting station. It is the number of degrees measured clockwise between the aircraft's heading and the direction from which the bearing is taken. The aircraft's magnetic heading (MH) is the direction the aircraft is pointed with respect to magnetic north. The magnetic bearing (MB) is the direction to or from a radio transmitting station measured relative to magnetic north.

ADF Indicator Instrument and Receiver.

NDB Components

The ground equipment, the NDB, transmits in the frequency range of 190 to 535 kHz. Most ADFs will also tune the AM broadcast band frequencies above the NDB band (550 to 1650 kHz). However, these frequencies are not approved for navigation because stations do not continuously identify themselves, and they are much more susceptible to sky wave propagation especially from dusk to dawn. NDB stations are capable of voice transmission and are often used for transmitting the automated weather observing system (AWOS). The aircraft must be in operational range of the NDB. Coverage depends on the strength of the transmitting station. Before relying on ADF indications, identify the station by listening to the Morse code identifier. NDB stations are usually two letters or an alpha-numeric combination.

ADF Components

The airborne equipment includes two antennas, a receiver, and the indicator instrument. The "sense" antenna (nondirectional) receives signals with nearly equal efficiency from all directions. The "loop" antenna receives signals better from two directions (bidirectional). When the loop and sense antenna inputs are processed together in the ADF radio, the result is the ability to receive a radio signal well in all directions but one, thus resolving all directional ambiguity. The indicator instrument can be one of four kind's fixedcard ADF, rotatable compass-card ADF, or radio magnetic indicator (RMI) with either one needle or dual needle. Fixed-card ADF (also known as the relative bearing indicator (RBI)) always indicates zero at the top of the instrument, with the needle indicating the RB to the station. Figure indicates an RB of 135°; if the MH is 045°, the MB to the station is 180° (MH + RB = MB to the station).

Relative bearing (RB) on a fixed-card indicator. Note that the card always indicates 360°, or north. In this case, the relative bearing to the station is 135° to the right. If the aircraft were on a magnetic heading of 360°, then the magnetic bearing (MB) would also be 135°.

The movable-card ADF allows the pilot to rotate the aircraft's present heading to the top of the instrument so that the head of the needle indicates MB to the station and the tail indicates MB from the station. Figure indicates a heading of 045°, MB to the station of 180°, and MB from the station of 360°.

Relative bearing (RB) on a movable-card indicator. By placing the aircraft's magnetic heading (MH) of 45° under the top index, the relative bearing (RB) of 135° to the right will also be the magnetic bearing (no wind conditions) which will take you to the transmitting station.

The RMI differs from the movable-card ADF in that it automatically rotates the azimuth card (remotely controlled by a gyrocompass) to represent aircraft heading. The RMI has two needles, which can be used to indicate navigation information from either the ADF or the VOR receiver. When a needle is being driven by the ADF, the head of the needle indicates the MB TO the station tuned on the ADF receiver. The tail of the needle is the bearing FROM the station. When a needle of the RMI is driven by a VOR receiver, the needle indicates where the aircraft is radially with respect to the VOR station. The needle points to the bearing TO the station, as read on the azimuth card. The tail of the needle points to the radial of the VOR the aircraft is currently on or crossing. Figure indicates a heading of 005°, the MB to the station is 015°, and the MB from the station is 195°.

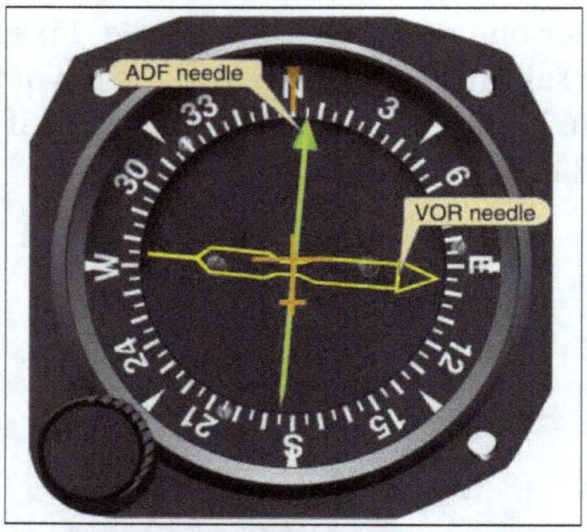

Radio magnetic indicator (RMI). Because the aircraft's magnetic heading is automatically changed, the relative bearing (RB), in this case 095°, will indicate the magnetic bearing (095°) to the station (no wind conditions) and the magnetic heading that will take you there.

Function of ADF

The ADF can be used to plot your position, track inbound and outbound, and intercept a bearing. These procedures are used to execute holding patterns and non-precision instrument approaches.

Orientation

The ADF needle points T_O the station, regardless of aircraft heading or position. The RB indicated is thus the angular relationship between the aircraft heading and the station, measured clockwise from the nose of the aircraft. Think of the nose/tail and left/right needle indications, visualizing the ADF dial in terms of the longitudinal axis of the aircraft. When the needle points to 0°, the nose of the aircraft points directly to the station; with the pointer on 210°, the station is 30° to the left of the tail; with the pointer on 090°, the station is off the right wingtip. The RB alone does not indicate aircraft position. The RB must be related to aircraft heading in order to determine direction to or from the station.

Station Passage

When you are near the station, slight deviations from the desired track result in large deflections of the needle. Therefore, it is important to establish the correct drift correction angle as soon as possible. Make small heading corrections (not over 5°) as soon as the needle shows a deviation from course, until it begins to rotate steadily toward a wingtip position or shows erratic left/right oscillations. You are abeam a station when the needle points 90° off your track. Hold your last corrected heading constant and time station passage when the needle shows either wingtip position or settles at or near the 180° position. The time interval from the first indications of station proximity to positive station passage varies with altitude—a few seconds at low levels to 3 minutes at high altitude.

Homing

The ADF may be used to "home" in on a station. Homing is flying the aircraft on any heading required to keep the needle pointing directly to the 0° RB position. To home in on a station, tune the station, identify the Morse code signal, and then turn the aircraft to bring the ADF azimuth needle to the 0° RB position. Turns should be made using the heading indicator. When the turn is complete, check the ADF needle and make small corrections as necessary.

Figure illustrates homing starting from an initial MH of 050° and an RB of 300°, indicating a 60° left turn is needed to produce an RB of zero. Turn left, rolling out at 50° minus 60° equals 350°. Small heading corrections are then made to zero the ADF needle.

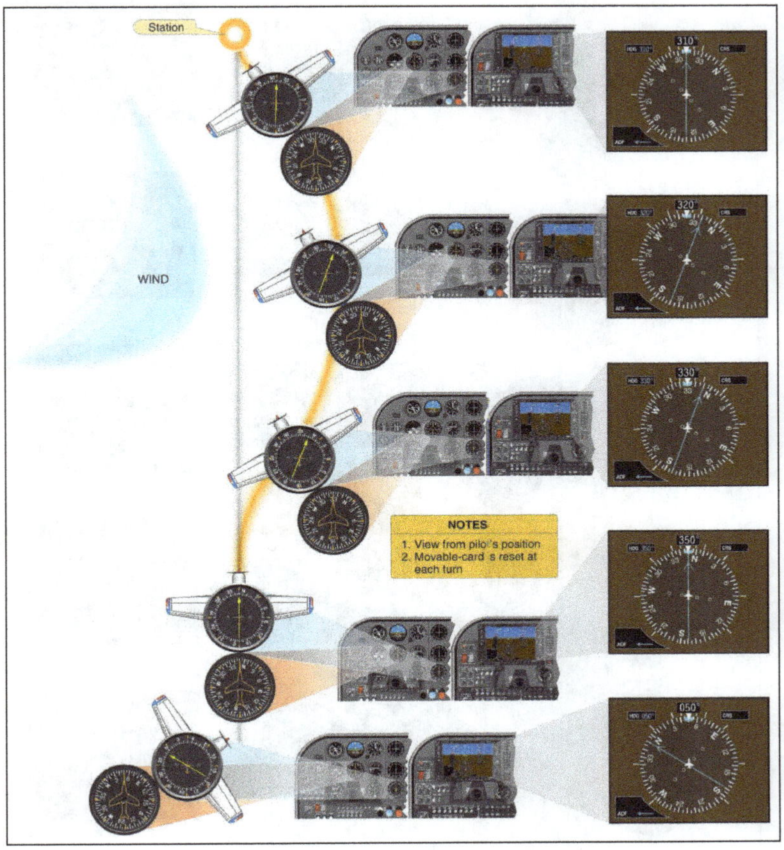

ADF Homing With a Crosswind.

If there is no wind, the aircraft will home to the station on a direct track over the ground. With a crosswind, the aircraft will follow a circuitous path to the station on the downwind side of the direct track to the station.

Tracking

Tracking uses a heading that will maintain the desired track to or from the station regardless of crosswind conditions. Interpretation of the heading indicator and needle is done to maintain a constant MB to or from the station.

To track inbound, turn to the heading that will produce a zero RB. Maintain this heading until off-course drift is indicated by displacement of the needle, which will occur if there is a crosswind (needle moving left = wind from the left; needle moving right = wind from the right). A rapid rate of bearing change with a constant heading indicates either a strong crosswind or close proximity to the station or both. When there is a definite (2° to 5°) change in needle reading, turn in the direction of needle deflection to intercept the initial MB. The angle of interception must be greater than the number of degrees of drift; otherwise the aircraft will slowly drift due to the wind pushing the aircraft. If repeated often enough, the track to the station will appear circular and the distance greatly increased as compared to a straight track. The intercept angle depends on the rate of drift, the aircraft speed, and station proximity. Initially, it is standard to double the RB when turning toward your course.

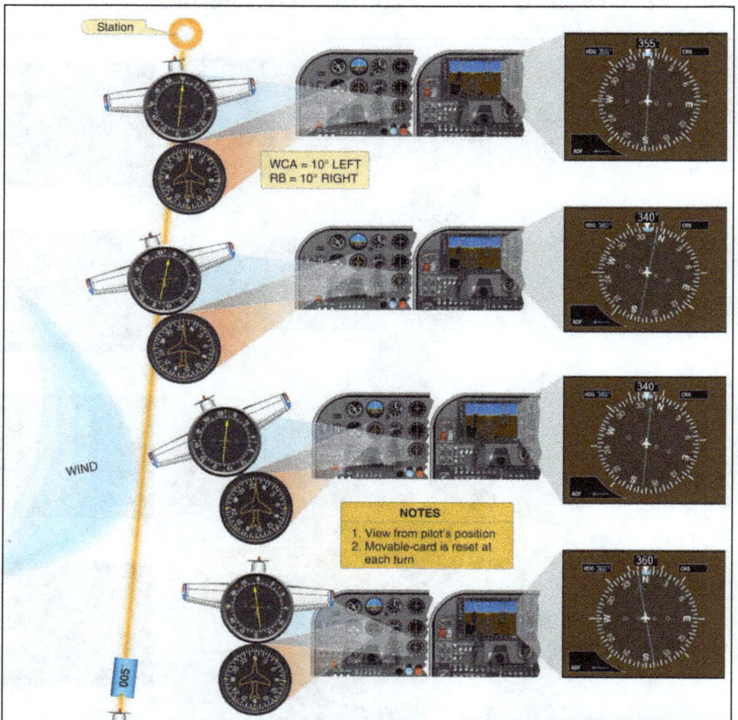

ADF Tracking Inbound.

For example, if you're heading equals your course and the needle points 10° left, turn 20° left, twice the initial RB. This will be your intercept angle to capture the RB. Hold this heading until the needle is deflected 20° in the opposite direction. That is, the deflection of the needle equals the interception angle (in this case 20°). The track has been intercepted, and the aircraft will remain on track as long as the RB remains the same number of degrees as the wind correction angle (WCA), the angle between the desired track and the heading of the aircraft necessary to keep the aircraft tracking over the desired track. Lead the interception to avoid overshooting the track. Turn 10° toward the inbound course. You are now inbound with a 10° left correction angle.

In figure, for the aircraft closest to the station, the WCA is 10° left and the RB is 10° right. If those values do not change, the aircraft will track directly to the station. If you observe off-course deflection in the original direction, turn again to the original interception heading. When the desired course has been re-intercepted, turn 5° toward the inbound course, proceeding inbound with a 15° drift correction. If the initial 10° drift correction is excessive, as shown by needle deflection away from the wind, turn to parallel the desired course and let the wind drift you back on course. When the needle is again zeroed, turn into the wind with a reduced drift correction angle.

To track outbound, the same principles apply needle moving left = wind from the left, needle moving right = wind from the right. Wind correction is made toward the needle deflection. The only exception is while the turn to establish the WCA is being made, the direction of the azimuth needle deflections is reversed. When tracking inbound, needle deflection decreases while turning to establish the WCA, and needle deflection increases when tracking outbound. Note the example of course interception and outbound tracking in figure.

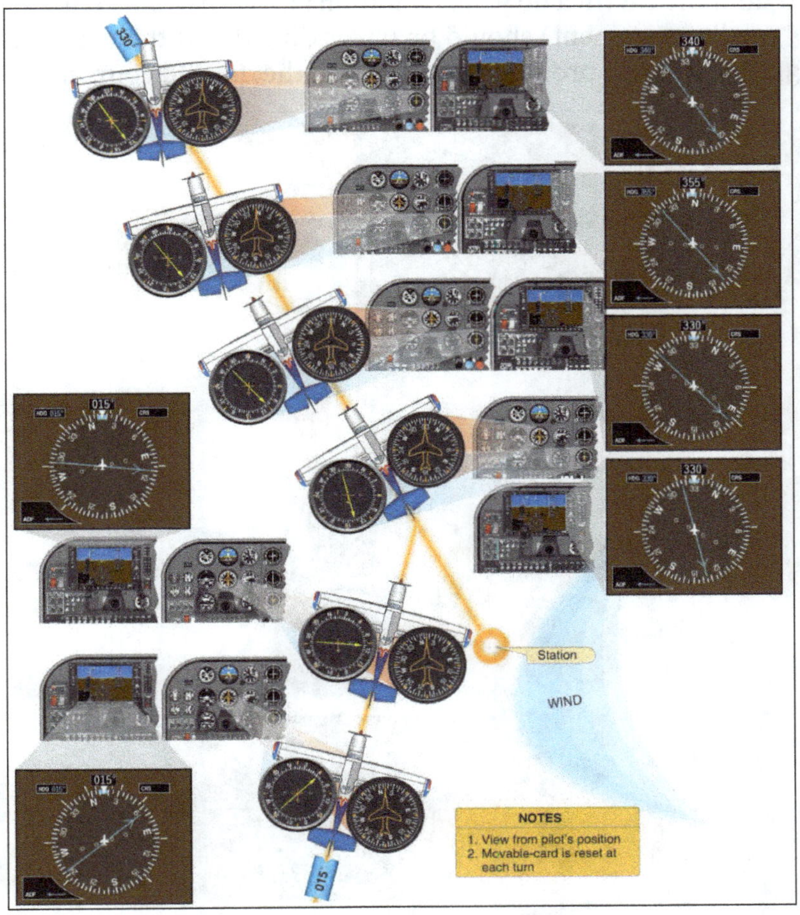

ADF Interception and Tracking Outbound.

Intercepting Bearings

ADF orientation and tracking procedures may be applied to intercept a specified inbound or outbound MB. To intercept an inbound bearing of 355°, the following steps may be used:

- Determine your position in relation to the station by paralleling the desired inbound bearing. In this case, turn to a heading of 355°. Note that the station is to the right front of the aircraft.
- Determine the number of degrees of needle deflection from the nose of the aircraft. In this case, the needle's RB from the aircraft's nose is 40° to the right. A rule of thumb for interception is to double this RB amount as an interception angle (80°).
- Turn the aircraft toward the desired MB the number of degrees determined for the interception angle which as indicated (in two above) is twice the initial RB (40°), or in this case 80°. Therefore, the right turn will be 80° from the initial MB of 355°, or a turn to 075° magnetic (355° + 80° + 075°).
- Maintain this interception heading of 075° until the needle is deflected the same number of degrees "left" from the zero position as the angle of interception 080°, (minus any lead appropriate for the rate at which the bearing is changing).
- Turn left 80° and the RB (in a no wind condition and with proper compensation for the rate of the ADF needle movement) should be 0°, or directly off the nose. Additionally, the MB should be 355° indicating proper interception of the desired course.

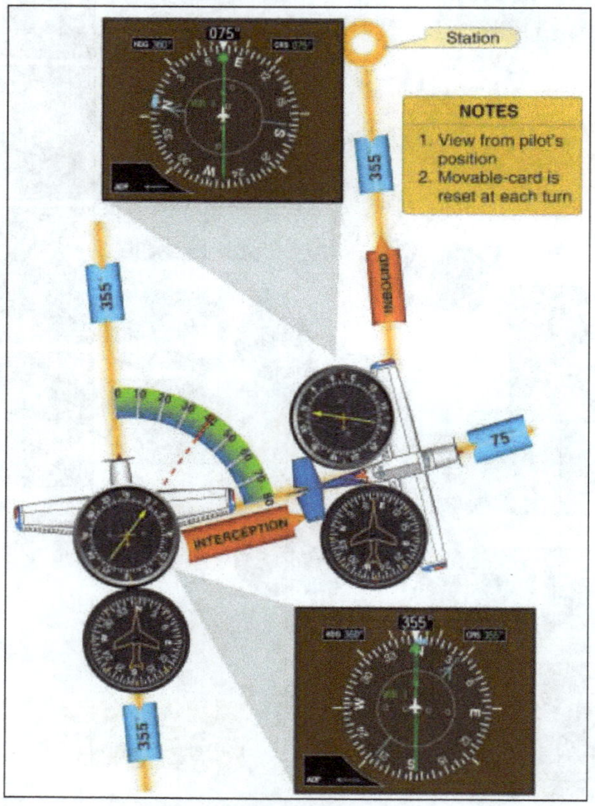

Interception of Bearing.

The rate of an ADF needle movement or any bearing pointer for that matter will be faster as aircraft position becomes closer to the station or waypoint (WP).

Interception of an outbound MB can be accomplished by the same procedures as for the inbound intercept, except that it is necessary to substitute the 180° position for the zero position on the needle.

Operational Errors of ADF

Some of the common pilot-induced errors associated with ADF navigation are listed below to help you avoid making the same mistakes. The errors are:

- Improper tuning and station identification. Many pilots have made the mistake of homing or tracking to the wrong station.
- Positively identifying any malfunctions of the RMI slaving system or ignoring the warning flag.
- Dependence on homing rather than proper tracking. This commonly results from sole reliance on the ADF indications, rather than correlating them with heading indications.
- Poor orientation, due to failure to follow proper steps in orientation and tracking.
- Careless interception angles, very likely to happen if you rush the initial orientation procedure.
- Overshooting and undershooting predetermined MBs, often due to forgetting the course interception angles used.
- Failure to maintain selected headings. Any heading change is accompanied by an ADF needle change. The instruments must be read in combination before any interpretation is made.
- Failure to understand the limitations of the ADF and the factors that affect its use.
- Over controlling track corrections close to the station (chasing the ADF needle), due to failure to understand or recognize station approach.
- Failure to keep the heading indicator set so it agrees with the magnetic compass.

Flight Control Systems

The architecture of the flight control system, essential for all flight operations, has significantly changed throughout the years. Nowadays two great categories of flight control systems can be found a full mechanical control on gliders and small general aviation, and a powered, or servo- assisted, control on large or combat aircraft.

One of the great additional effects after the introduction of servomechanisms is the possibility of using active control technology, working directly on the flight control actuators, for a series of benefits:

- Compensation for deficiencies in the aerodynamics of the basic airframe.
- Stabilisation and control of unstable airplanes, that have commonly higher performances.
- Flight at high angles of attack.
- Automatic stall and spinning protection.
- Gust alleviation.

A further evolution of the servo-assisted control is the fly-by-wire technique, based on signal processing of the pilot's demand before conversion into actuator control.

The number and type of aerodynamic surfaces to be controlled changes with aircraft category. Figure shows the classic layout for a conventional airliner. Aircraft have a number of different control surfaces those indicated in red form the primary flight control, i.e. pitch, roll and yaw control, basically obtained by deflection of elevators, ailerons and rudder (and combinations of them); those indicated in blue form the secondary flight control high-lift and lift-dump devices, airbrakes, tail trimming, etc.

Modern aircraft have often particular configurations, typically as follows:

- Elevons on delta wings, for pitch and roll control, if there is no horizontal tail.
- Flaperons, or trailing edge flaps-ailerons extended along the entire span.
- Tailerons, or stabillisers-ailerons (independently controlled).
- Swing wings, with an articulation that allows sweep angle variation.
- Canards, with additional pitch control and stabilisation.

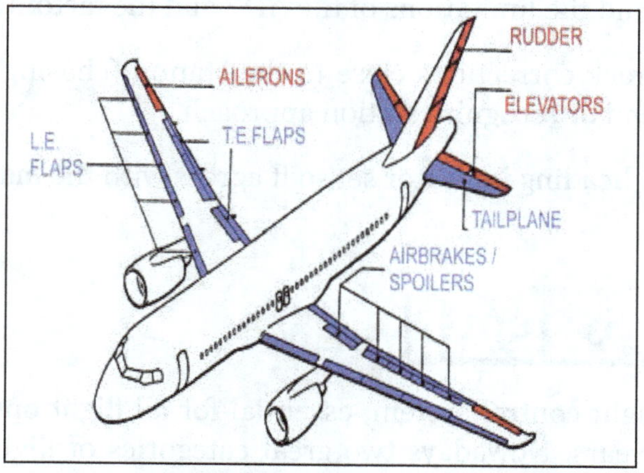

Flight control surfaces on airliner.

Primary flight control capability is essential for safety, and this aspect is dramatically emphasized in the modern unstable (military) airplanes, which could be not controlled without the continued operation of the primary flight control surfaces. For this reason the actuation system in charge of primary control has a high redundancy and reliability, and is capable of operating close to full performance after one or more failures.

Secondary actuation system failure can only introduce flight restriction, like a flapless landing or reduction in the max angle of attack; therefore it is not necessary to ensure full operation after failures.

Direct Mechanical Control

the linkage from cabin to control surface can be fully mechanical if the aircraft size and its flight envelop allow; in this case the hinge moment generated by the surface deflection is low enough to be easily contrasted by the muscular effort of the pilot.

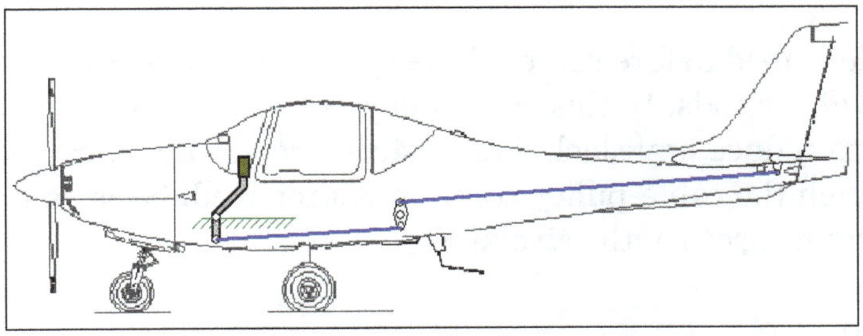

Push-pull rod system for elevator control.

Two types of mechanical systems are used push-pull rods and cable-pulley. In the first case a sequence of rods link the control surface to the cabin input. Bell-crank levers are used to change the direction of the rod routings figure sketches the push-pull control rod system between the elevator and the cabin control column; the bell-crank lever is here necessary to alter the direction of the transmission and to obtain the conventional coupling between stick movement and elevator deflection (column fwd = down deflection of surface and pitch down control).

From this simplified description the main requirements of a push-pull rod system are clear. First of all the linkage must be stiff, to avoid any unwanted deflection during flight and due to fuselage elasticity. Second, axial instability during compression must be excluded; the instability load P for a rod is given by:

$$P = \frac{\pi^2 EI}{\lambda^2}$$

Where,

E = Young modulus;

I = cross-section moment of inertia;

λ = reference length.

The reference length is linked to the real length of the rod, meaning that to increase the instability load the length must be decreased, or the rods must be frequently constrained by slide guides, or the routing must be interrupted with bell cranks.

Finally a modal analysis of the system layout is sometimes necessary, because vibrations of the rods can introduce oscillating deflections of the surface; this problem is particularly important on helicopters, because vibrations generated by the main rotor can induce a dramatic resonance of the flight control rods.

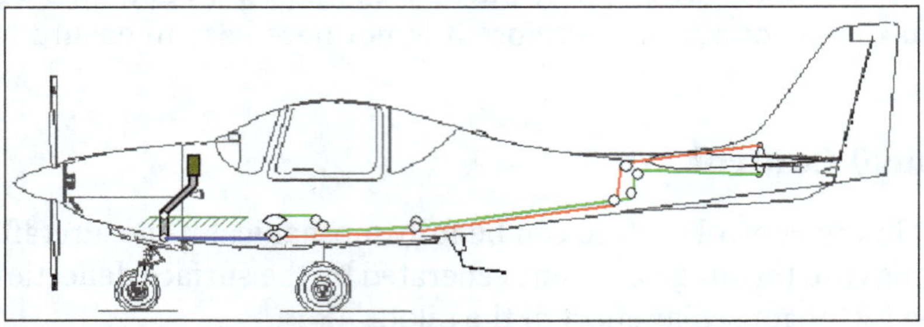

Cables and pulleys system for elevator control.

The same operation described before can be done by a cable-pulley system, where couples of cables are used in place of the rods. In this case pulleys are used to alter the direction of the lines, equipped with idlers to reduce any slack due to structure elasticity, cable strands relaxation or thermal expansion. Often the cable-pulley solution is preferred, because is more flexible and allows reaching more remote areas of the airplane.

Hydraulic Control

When the pilot's action is not directly sufficient for the control, the main option is a powered system that assists the pilot.

A few control surfaces on board are operated by electrical motors the hydraulic system has demonstrated to be a more suitable solution for actuation in terms of reliability, safety, weight per unit power and flexibility, with respect to the electrical system, then becoming the common tendency on most modern airplanes the pilot, via the cabin components, sends a signal, or demand, to a valve that opens ports through which high pressure hydraulic fluid flows and operates one or more actuators.

The valve, that is located near the actuators, can be signalled in two different ways mechanically or electrically; mechanical signalling is obtained by push-pull rods, or more commonly by cables and pulleys; electrical signalling is a solution of more modern and sophisticated vehicles .The basic principle of the hydraulic control is simple, but two aspects must be noticed when a powered control is introduced:

- The system must control the surface in a proportional way, i.e. the surface response (deflection) must be function to the pilot's demand (stick deflection, for instance).
- The pilot that with little effort acts on a control valve must have a feedback on the manoeuvre intensity.

The first problem is solved by using (hydraulic) servomechanisms, where the components are linked in such a way to introduce an actuator stroke proportional to the pilot's demand; many examples can be made, two of them are sketched in figure, the second one including also the hydraulic circuit necessary for a correct operation.

In both cases the control valve housing is solid with the cylinder and the cabin column has a mechanical linkage to drive the valve spool.

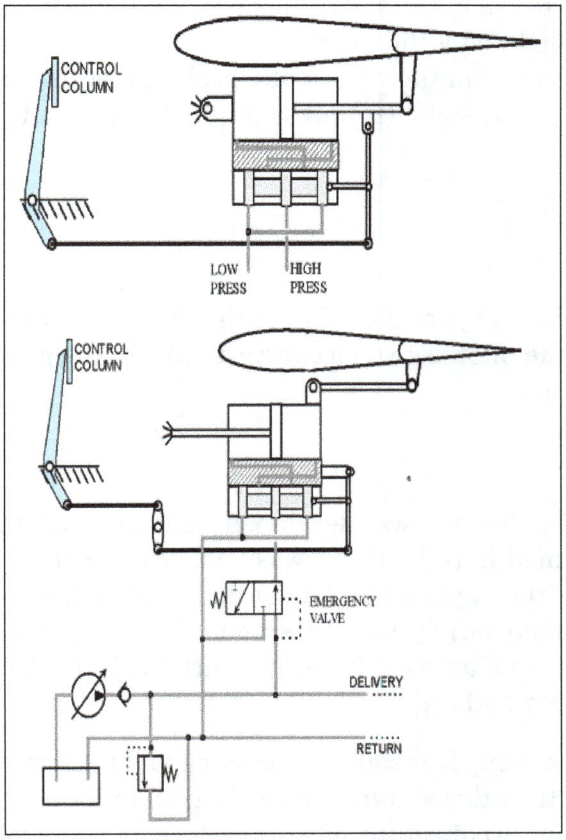

Classic hydraulic servomechanisms.

In the first case, the cylinder is hinged to the aircraft and, due to valve spool displacement and ports opening, the piston is moved in one direction or the other; the piston rod is also linked to the valve spool stick, in such a way that the piston movement brings the spool back towards its neutral position; when this is reached, the actuator stops, then obtaining a deflection that is proportional to the demand.

In the second case the piston is constrained to the aircraft; the cabin column controls the valve spool stick; this will result in a movement of the cylinder, and this brings the valve housing again towards the valve neutral position, then resulting in a stroke proportional to the pilot's demand. The hydraulic circuit also includes an emergency valve on the delivery segment to the control valve; if the delivery pressure drops, due for instance to a pump or engine failure, the emergency valve switches to the other position and links all the control valve inlets to the tank; this operation hydraulically unlocks the system, allowing the pilot for manual actuation of the cylinder.

It is clear now that the pilot, in normal hydraulic operating conditions, is requested for a very low effort, necessary to contrast the mechanical frictions of the linkage and the movement of the control valve the pilot is then no more aware of the load condition being imposed to the aircraft.

For this reason an artificial feel is introduced in powered systems, acting directly on the cabin control stick or pedals. The simplest solution is a spring system, then responding to the pilot's demand with a force proportional to the stick deflection; this solution has of course the limit to be not sensitive to the actual flight conditions. A more sophisticated artificial feel is the so-called Q feel. This system receives data from the pitot-static probes, reading the dynamic pressure, or the difference between total (p_t) and static (p_s) pressure, that is proportional to the aircraft speed v through the air density ρ:

$$p_t - p_s = \frac{1}{2}\rho v^2.$$

This signal is used to modulate a hydraulic cylinder that increases the stiffness in the artificial feel system, in such a way that the pilot is given a contrast force in the pedals or stick that is also proportional to the aircraft speed.

Fly-by-wire

In the 70's the fly-by-wire architecture was developed, starting as an analogue technique and later on, in most cases, transformed into digital. It was first developed for military aviation, where it is now a common solution; the supersonic Concorde can be considered a first and isolated civil aircraft equipped with a (analogue) fly-by-wire system, but in the 80's the digital technique was imported from military into civil aviation by Airbus, first with the A320, then followed by A319, A321, A330, A340, Boeing 777 and A380.

This architecture is based on computer signal processing and is schematically shown in figure the pilot's demand is first of all transduced into electrical signal in the cabin and sent to a group of independent computers (Airbus architecture substitute the cabin control column with a side stick); the computers sample also data concerning the flight conditions and servo-valves and actuators positions; the pilot's demand is then processed and sent to the actuator, properly tailored to the actual flight status. The flight data used by the system mainly depend on the aircraft category; in general the following data are sampled and processed:

- Pitch, roll, yaw rate and linear accelerations.
- Angle of attack and sideslip.
- Airspeed/mach number, pressure altitude and radio altimeter indications.
- Stick and pedal demands.
- Other cabin commands such as landing gear condition, thrust lever position, etc.

The full system has high redundancy to restore the level of reliability of a mechanical or hydraulic system, in the form of multiple (triplex or quadruplex) parallel and independent lanes to generate and transmit the signals, and independent computers that process them; in many cases both hardware and software are different, to make the generation of a common error extremely remote, increase fault tolerance and isolation; in some cases the multiplexing of the digital computing and signal transmission is supported with an analogue or mechanical back-up system, to achieve adequate system reliability.

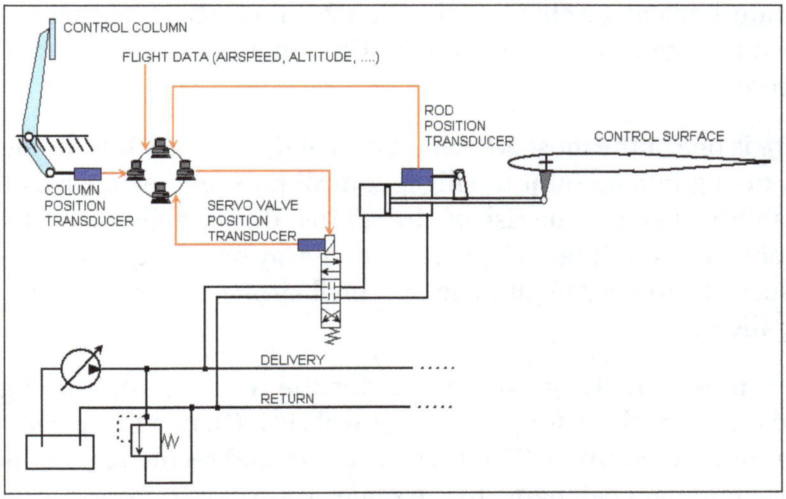

Fly-by-wire system.

The benefits of the fly-by-wire architecture are different, and vary significantly between military and civil aircraft; some of the most important benefits are as follows:

- Flight envelope protection (the computers will reject and tune pilot's demands that might exceed the airframe load factors).
- Increase of stability and handling qualities across the full flight envelope, including the possibility of flying unstable vehicles.
- Turbulence suppression and consequent decrease of fatigue loads and increase of passenger comfort.
- Use of thrust vectoring to augment or replace lift aerodynamic control, then extending the aircraft flight envelope.
- Drag reduction by an optimised trim setting.
- Higher stability during release of tanks and weapons.
- Easier interfacing to auto-pilot and other automatic flight control systems.
- Weight reduction (mechanical linkages are substituted by wirings).
- Maintenance reduction.
- Reduction of airlines' pilot training costs (flight handling becomes very similar in a whole aircraft family).

For civil fly-by-wire aircraft in normal operation the flight control changes according to the flight mode ground, take-off, flight and flare. Transition between modes is smooth and the pilot is not affected in its ability to control the aircraft in ground mode the pilot has control on the nose wheel steering as a function of speed, after lift-off the envelope protection is gradually introduced and in flight mode the aircraft is fully protected by exceeding the maximum negative and positive load factors (with and without high lift devices extracted), angle of attack, stall, airspeed/Mach number, pitch attitude, roll rate, bank angle etc.; finally, when the aircraft approaches to ground the control is gradually switched to flare mode, where automatic trim is deactivated and modified flight laws are used for pitch control.

The control software is one of the most critical aspects of fly-by-wire. It is developed in accordance to very strict rules, taking into account the flight control laws, and extensive testing is performed to reduce the probability of error. The risk of aircraft loss due to flight control failure is 2×10^{-6} per flight hour for a sophisticated military airplane that anyway has the ejection seat as ultimate solution; the risk is reduced to 10^{-9} per flight hour for a civil airplane, were occupants cannot evacuate the airplane during flight.

Figure shows, as example, the fly-by-wire layout for the Airbus 340. Three groups of personal computers are used on board three for primary control (FCPC), two for secondary control (FCSC) and two for high lift devices control (SFCC). The primary and secondary computers are based on different hardware; computers belonging to the same group have different software.

Two additional personal computers are used to store flight data.

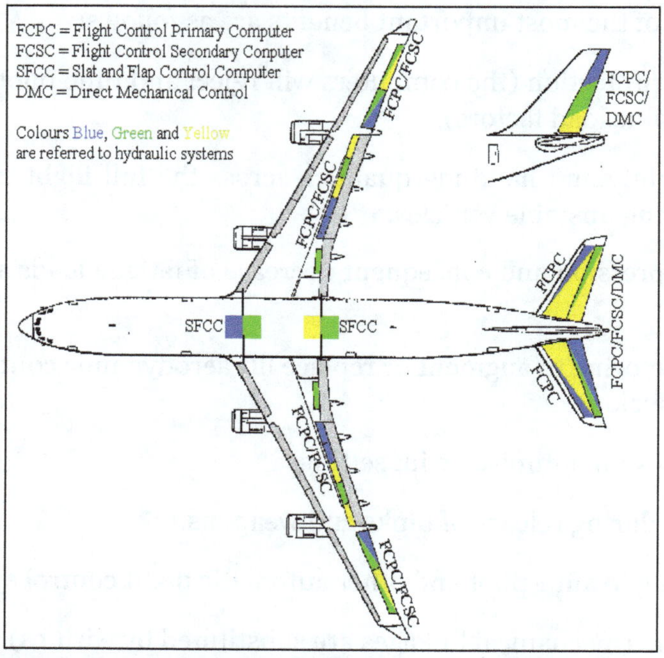

A340 fly-by-wire layout, including hydraulic system indications.

In the drawing the computer group and hydraulic system that control each surface are indicated (there are three independent hydraulic systems on the A340, commonly indicated as Blue, Yellow and Green). The leading edge flaps are linked together, and so are the trailing edge flaps, and then they are controlled by hydraulic units in the fuselage.

The drawing shows a considerable redundancy of the flight control system the inboard and outboard ailerons, elevators and rudder are controlled by both the primary and secondary computers and operated by the three hydraulic sub-systems; the high lift devices are controlled by their specific computers and operated by the three hydraulic systems (Blue and Green for the leading edge, Yellow and Green for the trailing edge); the vertical stabiliser, having a secondary role, is controlled only by the secondary computers and operated by two hydraulic sub-systems. Thanks to this layout, first of all, in case of double hydraulic sub-system fault, the aircraft can be basically controlled with one hydraulic sub-system. Moreover, in case of total power black-out, the pilot can control the rudder and elevators by a mechanical back-up system, since the capability of this aircraft to land safely has been demonstrated with only limited pitch and yaw control.

Fly-by-wire architecture is inevitable for some aircraft categories figure shows a typically unstable aircraft and a tilt rotor aircraft.

Northrop B-2.

Bell-Boeing V-22: Need of fly-y-wire architecture for unstable (B-2) and thrust vectoring (V-22) airplanes.

The first example is the Northrop B-2, a stealth bomber; this airframe is designed with the primary aim of low observability, thanks to a limited radar wave reflection; this need brought to an external shape with particular aerodynamic surfaces, and then the necessity of a fly-by-wire system for flight control.

In the second case, the Bell-Boeing V-22, the peculiarity is the thrust vectoring through complete rotation of the rotor-engine assembly, enabling the aircraft to vertical takeoff and landing, like a helicopter, but normal fixed wing horizontal flight. Transition between take-off/landing manoeuvres and horizontal flight would require an uncommon expertise from the pilots, but is easily managed by a fly-by-wire architecture.

Actuation Systems

As observed, the current tendency in flight control is to install hydraulic actuators and mechanically or electrically signalled control valves to operate the actuators. Figure shows an example of electrically signalled valve that controls a four-chamber actuator, powered by two hydraulic sub-systems; the layout is such that, in case of one hydraulic sub-system fault, this is automatically excluded and the actuator is operated by the other functioning sub-system. In case of complete hydraulic fault, the actuator is locked to current position.

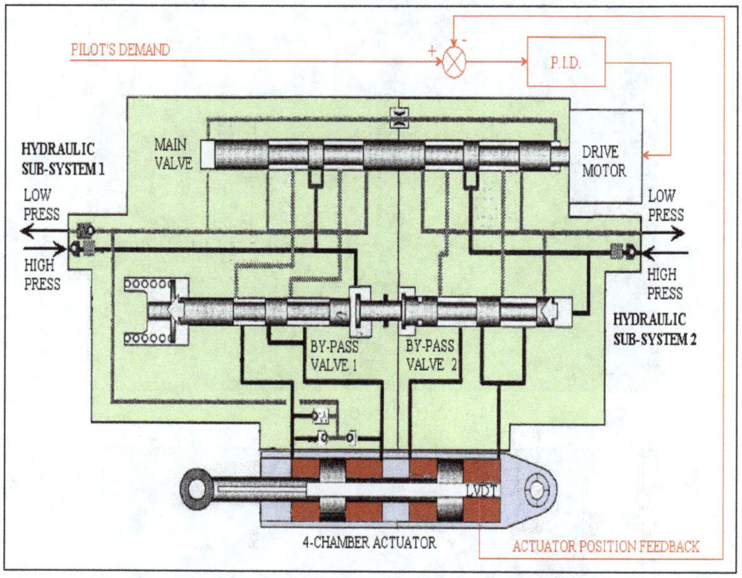

Electrical servo-valve integrated in actuator.

The actuator position is measured by LVDT's (displacement transducers); the pilot's demand is compared with the LVDT's feedback and the result is sent to the PID (proportional-integrative-derivative) control. This is actually a simplified description of the control circuit, because it usually includes amplifiers, filters and often a feedback from the main valve LVDT, which is roughly proportional to the piston velocity. The most important parameters that define actuators performances are stall load, maximum rate capability, frequency response, dynamic stiffness and failure transients.

The stall load is the maximum force on the piston that can be contrasted by the hydraulic pressure, in both directions; this basic parameter is affected by piston area, hydraulic pressure and related head losses.

Piston rates are defined in both directions and are an important requirement because they are necessary to define flow rates and components sizing. Of course they also depend on the applied load, up to the stall load when the rate is zero; a common practice is to indicate piston rates at zero-load and 70% stall load condition. Frequency response requirements are in terms of gain and phase-lag boundaries of the displacement under representative loadings.

Dynamic stiffness, or impedance, is the ability of actuator to resist an external oscillatory load, in terms of stiffness and damping requirements on the impedance are aimed at avoiding flutter for the control surface. Failure transient is the piston displacement after failure, as a function of time, in percent of the full stroke.

Up to now only hydraulic actuators have been considered. Two different types of alternatives that are under study are the electro-mechanical actuator and the electro hydrostatic actuator, EMA and EHA respectively. The first system is made of an electrical motor and gearbox assembly that moves the ram. These types of actuators have been used on aircraft for decades for trims, doors operation and other uses that require low power, low motive force and low response times. At present possible development of this system and extension to flight control is made possible by high voltage DC motors (270 VDC), high power solid-state switches and control microprocessors. Studies are currently concentrated to make the electro-mechanical actuators as light and reliable as the corresponding electro-hydraulic ones.

A second actuation system under development is the electro-hydrostatic actuator. This is made of a self-contained hydraulic system, including pump and reservoir, which moves a piston, and an electric motor that drives the pump. The system is activated only during the demand; otherwise it is locked in the fixed position. Also in this case the development of the system is made possible by the use of a 270 VDC electrical system.

Helicopter Flight Control System

Helicopter systems show significant differences with respect to fixed-wing systems, especially as far as the flight control is concerned. Rotary wing aircraft are more complex that fixed wings they are basically non-symmetric machines, with strong interactions among the flight controls.

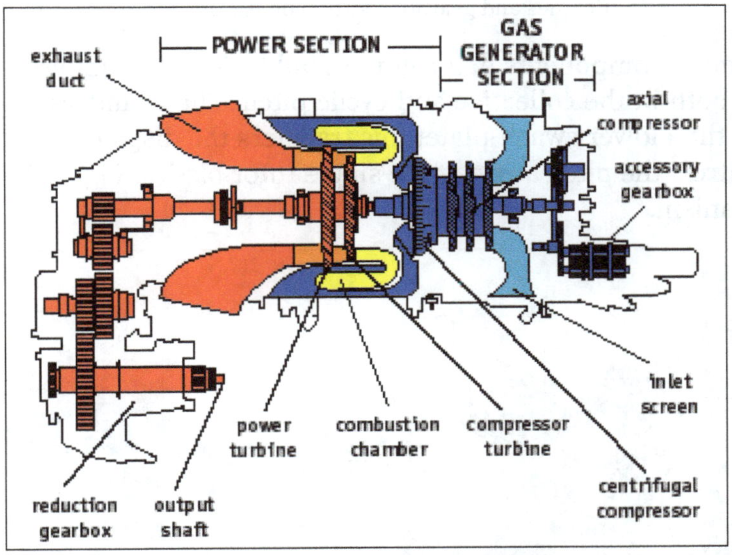

Turbo-shaft engine for helicopters.

With exception of small helicopters, which are equipped with piston engines, turbo shaft propulsors are used to power the main and tail rotor. This means that almost all the power developed is converted into mechanical energy, also including driving of hydraulic and electrical generators the exhaust gases have neglectable thrust power. For this reason the engine has a double turbine, as schematically shown in fig the first one (compressor turbine), after the combustion chamber, drives the shaft with the compressor; the second one (power turbine) drives the output gearbox shaft that power the helicopter rotors; this is a free turbine, independent from the compressor turbine, so that the engine has little inertia and transients and can spin up rapidly to its functioning regime; the power turbine, driving the rotors, has high inertia and spins up slowly. Helicopters may have two or more engines, rarely single engine. They are located on the cabin roof, close to the main rotor gearbox, as shown in the example of figure.

Flight control on helicopters is obtained by changing the configuration of the main and tail rotors. The lift force is generated by the blades of the main rotor that are usually in a number between two and six. By changing their pitch (collective pitch), their lift coefficient is increased. This has a consequence of increasing the torque on the main shaft that figure must be compensated by changing the pitch of the tail rotor to keep the vehicle in a steady heading. The control in longitudinal and sideward motion is obtained by tilting the main rotor in the corresponding direction (cyclic pitch).

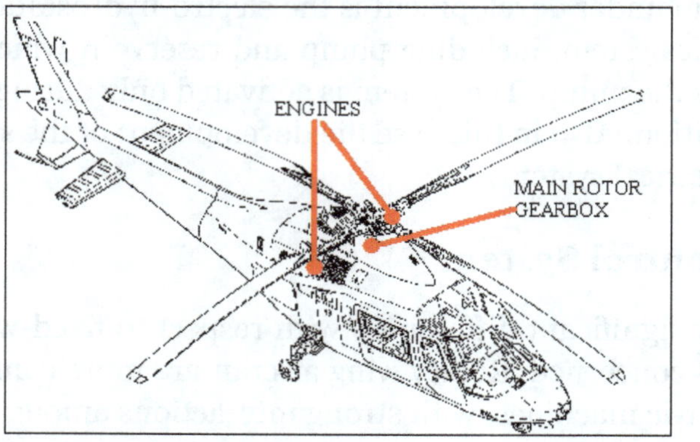

Engines and gearbox location on Agusta A109.

Figure A shows the main components of a rotor assembly. Three cylinders are enough to control the rotor operation, both in the collective and cyclic pitch. The cylinders, signalled by the pilot's demand, displace or tilt a lower swash-plate; this transfers the displacement to an upper rotating swash-plate that controls the pitch angle of the single rotor blades. Figure B helps in the comprehension of the mechanism.

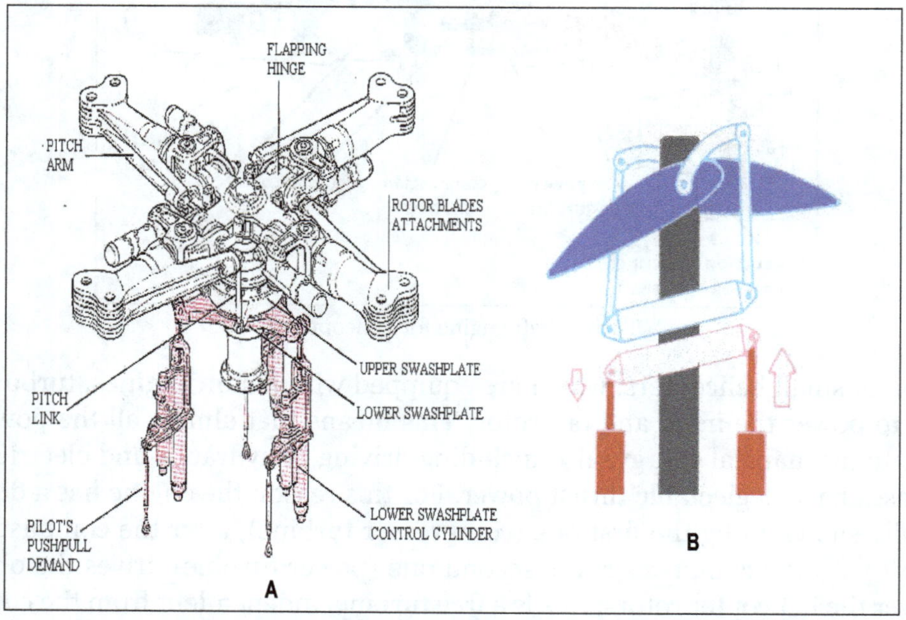

Main rotor control components and pitch control schematic description.

The change in pitch angle of the blades changes the blade lift; the collective pitch, controlled through the collective pitch lever on the pilot's side, operates simultaneously on all the blades, resulting in a total lift increase of the vehicle; the cyclic pitch, controlled by the pilot's stick, will unbalance the lift distribution on the rotor disc and will bring it to tilt, then generating a horizontal force component. Flapping hinges allow moderate tilting of the blades, due to the fact that the advancing blade has a relative airflow velocity that can be considerably higher than the backing blade, then being subject to a higher lift force.

Yaw control, which consists in changing the pitch of the tail rotor blades, is finally obtained through pedals, like on fixed wing aircraft. Engine power is controlled by a throttle twist-grip on the collective lever this must be operated to compensate all pitch variations, because they affect the rotor shaft torque and the lift orientation; therefore helicopter control is a complex co-ordination of collective/cyclic pitch, tail lateral pitch and engine power control.

From this rough description it becomes clear that the helicopter is far from being a stable vehicle flying a helicopter is more difficult than flying a fixed wing aircraft and this gap is emphasized during manoeuvre and flight in turbulence. For this reason modern advanced helicopters are equipped with automatic engine control, flight auto-stabilisation and autopilot systems, capable of minimising the effect of interactions among the controls, and then reducing the pilot's workload.

References

- Pitot-Static-System: skybrary.aero, Retrieved 19 May, 2019
- Airspeed-indicator, avionics-and-instruments, notebook: cfinotebook.net, Retrieved 28 April, 2019
- Pitot-static-pressure-sensing-flight-instruments-part-two: flight-mechanic.com, Retrieved 21 July, 2019
- Altimeter, avionics-and-instruments, notebook: cfinotebook.net, Retrieved 7 May, 2019
- Compass-systems-magnetic-compass: flight-mechanic.com, Retrieved 15 January, 2019
- Compass-systems-the-vertical-card-magnetic-compass: flight-mechanic.com, Retrieved 1 June, 2019
- Gyromagnetic-Compass: skybrary.aero, Retrieved 1 August, 2019
- Gyroscopic-flight-instruments-part-one: flightliteracy.com, Retrieved 9 May, 2019
- Very-high-frequency-omnidirectional-range-vor-part-one: flightlearnings.com, Retrieved 8 August , 2019

Permissions

All chapters in this book are published with permission under the Creative Commons Attribution Share Alike License or equivalent. Every chapter published in this book has been scrutinized by our experts. Their significance has been extensively debated. The topics covered herein carry significant information for a comprehensive understanding. They may even be implemented as practical applications or may be referred to as a beginning point for further studies.

We would like to thank the editorial team for lending their expertise to make the book truly unique. They have played a crucial role in the development of this book. Without their invaluable contributions this book wouldn't have been possible. They have made vital efforts to compile up to date information on the varied aspects of this subject to make this book a valuable addition to the collection of many professionals and students.

This book was conceptualized with the vision of imparting up-to-date and integrated information in this field. To ensure the same, a matchless editorial board was set up. Every individual on the board went through rigorous rounds of assessment to prove their worth. After which they invested a large part of their time researching and compiling the most relevant data for our readers.

The editorial board has been involved in producing this book since its inception. They have spent rigorous hours researching and exploring the diverse topics which have resulted in the successful publishing of this book. They have passed on their knowledge of decades through this book. To expedite this challenging task, the publisher supported the team at every step. A small team of assistant editors was also appointed to further simplify the editing procedure and attain best results for the readers.

Apart from the editorial board, the designing team has also invested a significant amount of their time in understanding the subject and creating the most relevant covers. They scrutinized every image to scout for the most suitable representation of the subject and create an appropriate cover for the book.

The publishing team has been an ardent support to the editorial, designing and production team. Their endless efforts to recruit the best for this project, has resulted in the accomplishment of this book. They are a veteran in the field of academics and their pool of knowledge is as vast as their experience in printing. Their expertise and guidance has proved useful at every step. Their uncompromising quality standards have made this book an exceptional effort. Their encouragement from time to time has been an inspiration for everyone.

The publisher and the editorial board hope that this book will prove to be a valuable piece of knowledge for students, practitioners and scholars across the globe.

Index

www.ingramcontent.com/pod-product-compliance
Lightning Source LLC
LaVergne TN
LVHW080647210926
848485LV00003B/342